Principles of
Measurement and Instrumentation

Principles of
Measurement and Instrumentation

Alan S. Morris

Department of Control Engineering
University of Sheffield

PRENTICE HALL
New York London Toronto Sydney Tokyo

First published 1988 by
Prentice Hall International (UK) Ltd,
66 Wood Lane End, Hemel Hempstead,
Hertfordshire, HP2 4RG

A division of
Simon & Schuster International Group

Printed and bound in Great Britain at the
University Press, Cambridge

Library of Congress Cataloging-in-Publication Data

Morris, Alan S., 1948-
  Principles of measurement and instrumentation /
Alan S. Morris
    p.   cm
  Bibliography: p.
  Includes index.
  ISBN 01-13-710179-1
  1. Engineering instruments.   2. Engineering—
Measurement.
I.   Title.
TA165. M638 1988
681.2 — dc19   87-29266
                                                    CIP

---

1 2 3 4 5 92 91 90 89 88

ISBN 0-13-710179-1
ISBN 0-13-710161-9 PBK

To Jane, Nicola and Julia

Contents

Preface

Measurement has been of great relevance to mankind since the earliest days of human civilization, where it was first used as a means of quantifying the exchange of goods in barter trade systems. Today, measurement systems, and the instruments and transducers used in them, are of immense importance in a wide variety of domestic and industrial activities. The growth in the number and sophistication of instruments used in industry has been particularly significant over the last two decades as automation schemes have been developed. A similar rapid expansion in their use has also been evident in military and medical applications over the same period.

The aim of this book is to present the subject of instrumentation, and its use within measurement systems, as an integrated and coherent subject. The text is divided into two parts, the first covering the principles of measurement, and the second presenting the range of instruments available for measuring various physical quantities. This order of coverage has been chosen so that the general characteristics of measuring instruments and their behavior in different operating environments are well established before the reader is introduced to the various types of instrument available. This method of approach means that the reader will have gained a thorough understanding of the theoretical considerations which govern the choice of a suitable instrument in any particular measurement situation before the properties of any particular instrument are discussed, and so will be properly equipped to appreciate and critically appraise the various merits and characteristics of the ranges of instrument presented.

Part A starts with a general introduction to the static and dynamic characteristics of measuring instruments. This is immediately followed by a discussion about the computational components of intelligent instruments, in recognition of the growing importance of these in modern instrumentation technology. This order of treatment enables the contribution of intelligence in instruments toward reducing measurement errors to be developed throughout the following chapters in part A, and indeed throughout the rest of the book. These other chapters in part A are concerned mainly with examining the sources and avoidance of measurement error, and considering the statistical analysis of instrument output data. In recognition of their importance in measuring the resistance-, capacitance- or inductance-change type of output associated with many instruments, a whole chapter is devoted in part A to the analysis of bridge circuits. Finally, part A concludes with a presentation of the instruments available for measuring and recording the electrical outputs from instruments. Throughout the first part of the book, various exercises are set to allow the reader to test his/her understanding of the material presented. Self-checking is provided for by a set of solutions contained in an appendix at the end of the book.

Part B of the book consists of 10 chapters, which are concerned with presenting the ranges of instruments available for measuring the major physical quantities arising within industrial processes. Apart from presenting the main instruments in use in each case, coverage is also given to many less common instruments used in special applications, and particularly to some recently developed instruments which are likely to gain more widespread use in the future. Some general guidance is given within each chapter about how to go about choosing a suitable instrument for any particular measurement situation.

Whilst the motivation for producing this book has been to provide a supporting text for introductory courses in measurement and instrumentation, the depth of coverage in many areas extends beyond the minimum level required for introductory courses and enables the book to be also used in support of courses which are some-what more advanced. The amount of mathematical notation within the text has been minimized as far as possible to make the book suitable as a course text at all levels of further and higher education.

Besides this role as a student course text, the book is also of relevance to the practicing instrumentation technologist who wishes to upgrade his/her knowledge about the latest developments in the theory of measurement principles. The catalog of instrument ranges contained in part B is particularly recommended as a valuable source of reference to all practicing engineers whose duties involve the design, specification or maintenance of instruments.

A.S.S.

PRINCIPLES OF MEASUREMENT

1

Introduction to Measurement

The last decade has seen a large and rapid growth in new industrial technology, encouraged by developments in electronics in general and computers in particular. This decade of rapid growth, often referred to as the 'electronics revolution', represents a quantum leap forward in production techniques of a similar magnitude to that brought about by the Industrial Revolution in the last century. At the spearhead of this thrust forward has been the digital computer, this being an essential component in both hard production automation schemes and in the even more advanced flexible manufacturing systems which are just beginning to emerge.

The massive growth in the application of computers to process control and monitoring tasks has spawned a parallel growth in the requirement for instruments to measure, record and control process variables. As modern production techniques dictate working to tighter and tighter accuracy limits, and as economic forces limiting production costs become more severe, so the requirement for instruments to be both accurate and cheap becomes ever harder to satisfy. This latter problem is at the focal point of the research and development efforts of all instrument manufacturers. In the past few years the most cost-effective means of improving instrument accuracy has been found in many cases to be the inclusion of digital computing power within instruments themselves. These intelligent instruments therefore feature prominently in current instrument manufacturers' catalogs.

Such intelligent instruments represent the latest stage in the present era of measurement technology. This era goes back to the start of the Industrial Revolution in the nineteenth century when measuring instruments first began to be developed to satisfy the needs of industrialized production techniques. However, the complete history of measurement techniques goes back much further, in fact by thousands of years to the very start of human civilization. As mankind evolved from his ape-like ancestors, he ceased to rely on using caves for shelter and to forage whatever he could for food, and started instead to build his own shelters and produce food by planting seed, rearing animals and farming in an organized manner. Initially, civilized mankind lived in family communities and one assumes that such groupings were able to live in reasonable harmony without great arguments about who was working hardest and who was consuming the most. With the natural diversity of human talents, however, particular family groups developed particular specializations. Some communities would excel in farming, perhaps because of the arable quality of the area of land they occupied, whilst other groups might be particularly proficient at building houses. This inevitably led to family communities producing an excess of some things and a deficit in others. Trade of excess production between family communities therefore developed naturally, and followed a barter system whereby produce or work of one sort would be exchanged for produce or work of another.

Clearly, this required a system of measurement to quantify the amounts being exchanged and to establish clear rules about the relative values of different commodities. Such early systems of measurement were based on whatever was available as a measuring unit. For purposes of measuring length, the human torso was a convenient tool, and gave us units of the hand, the foot and the cubit. Although generally adequate for barter trade systems, such measurement units are of course imprecise, varying as they do from one person to the next, and modern measurement systems are therefore based on units which are defined much more accurately.

Today, the techniques of measurement are of immense importance in most facets of human civilization. Present-day applications of measuring instruments can be classified into three major areas. The first of these is their use in regulating trade, applying instruments which measure physical quantities such as length, volume and mass in terms of standard units. The particular instruments and transducers employed in such applications are included in the general description of instruments presented in part B of this book.

The second application area of measuring instruments is in monitoring functions. These provide information which enables human beings to take some prescribed action accordingly. The gardener uses a thermometer to determine whether he should turn the heat on in his greenhouse or open the windows if it is too hot. Regular study of a barometer allows us to decide whether we should take our umbrellas if we are planning to go out for a few hours. Whilst there are thus many uses of instrumentation in our normal domestic lives, a considerable majority of monitoring functions exist to provide the information necessary to allow a human being to

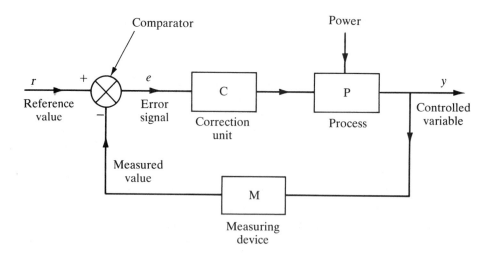

Figure 1.1   Elements of a simple closed-loop control system

control some industrial operation or process. In a chemical process, for instance, the progress of chemical reactions is indicated by the measurement of temperatures and pressures at various points, and such measurements allow the operator to take correct decisions regarding the electrical supply to heaters, cooling water flows, valve positions, etc. One other important use of monitoring instruments is in calibrating the instruments used in the automatic process control systems described below.

Use as part of automatic control systems forms the third application area of measurement systems. Figure 1.1 shows a functional block diagram of a simple control system where some output variable y of a controlled process P is maintained at a reference value r. The value of the controlled variable y, as measured by a measuring instrument M, is compared with the reference value r, and the difference e is applied as an error signal to the correcting unit C. The correcting unit then modifies the process output such that the output variable is given by $y=r$. A simple example of such a process is a shunt-wound d.c. servomotor with fixed field excitation as shown in figure 1.2. The output variable requiring control is the motor speed, and this is achieved by varying the armature output according to the difference between the measured speed and the demanded reference speed r. The characteristics of measuring instruments in such feedback control systems are of fundamental importance to the quality of control achieved. The accuracy and resolution with which an output variable of a process is controlled can never be better than the accuracy and resolution of the measuring instruments used. This is a very important principle, but one which is often inadequately discussed in many texts on automatic control systems. Such texts explore the theoretical aspects of control system design in considerable depth, but fail to give sufficient emphasis to the fact that all gain and phase margin performance calculations, etc., are entirely dependent on the quality of the process measurements obtained.

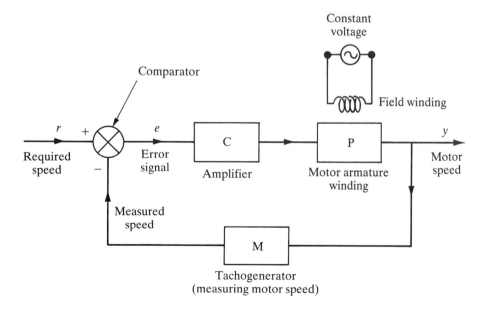

Figure 1.2   Control of a shunt-wound d.c. servomotor

A measuring instrument exists to provide information about the physical value of some variable being measured. In simple cases, an instrument consists of a single unit which gives an output reading or signal according to the magnitude of the unknown variable applied to it. In more complex measurement situations, however, a measuring instrument may consist of several separate elements as shown in figure 1.3. These components might be contained within one or more boxes, and the boxes holding individual measurement elements might be either close together or physically separate. Because of this modular nature of the elements within it, it is common to refer to a measuring instrument as a measurement system, and this latter term is used extensively throughout this book to emphasize the modular nature of measuring instruments.

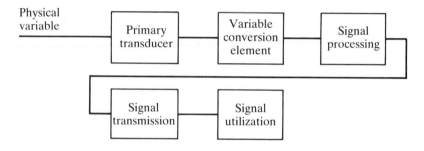

Figure 1.3   Elements of a measuring instrument

Common to any measuring instrument is the primary transducer: this gives an output which is a function of the measurand (the input applied to it). For most, but not all, transducers, this function is at least approximately linear. Some examples of primary transducers are a liquid-in-glass thermometer, a thermocouple and a strain gauge. In the case of the mercury-in-glass thermometer, the output reading is given in terms of the level of the mercury, and so this particular primary transducer is also a complete measurement system in itself. In general, however, the primary transducer is only part of a measurement system. The range of primary transducers available for measuring a wide range of physical quantities is presented in part B of this book.

The output variable of a primary transducer is often in an inconvenient form and has to be converted to a more convenient one. The displacement-measuring strain gauge, for instance, has an output in the form of a varying resistance. This is converted to a change in voltage by a *bridge circuit*, which is a typical example of the variable conversion element shown in figure 1.3.

Signal processing elements exist to improve the quality of the output of a measurement system in some way. A very common type of signal processing element is the electronic amplifier which amplifies the output of the primary transducer or variable conversion element, thus improving the sensitivity and resolution of measurement. This element of a measuring system is particularly important where the primary transducer has a low output. Thermocouples, for instance, have a typical output of only a few millivolts. Other types of signal processing element are those which filter out induced noise and remove mean levels, etc.

The observation or application point of the output of a measurement system is often some physical distance away from the site of a primary transducer which is measuring a physical quantity, and some mechanism of transmitting the measured signal between these points is necessary. Sometimes, such separation is made solely for purposes of convenience, but more often it follows from the physical inaccessibility or environmental unsuitability of the site of the primary transducer for mounting the signal presentation/recording unit. The signal transmission element traditionally has consisted of single- or multi-cored cable, which is often screened to minimize signal corruption by induced electrical noise. Optical fiber cables are, however, being used in ever-increasing numbers in modern installations, in part because of their low transmission loss and imperviousness to the effects of electrical and magnetic fields.

The final element in a measurement system is the point where the measured signal is utilized. In some cases, this element is omitted altogether because the measurement is used as part of an automatic control scheme and the transmitted signal is fed directly into the control system. In other cases, this element in the measurement system takes the form either of a signal presentation unit or of a signal recording unit. These take many forms according to the requirements of the particular measurement application, and the range of possible units is discussed more fully in chapter 8.

2

Instrument Classification and Characteristics

2.1 Instrument classification

Instruments can be subdivided into separate classes according to several criteria. These subclassifications are useful in broadly establishing several attributes of particular instruments such as accuracy, cost, and general applicability to different applications.

Active/passive instruments

Instruments are divided into active or passive ones according to whether the instrument output is entirely produced by the quantity being measured or whether the quantity being measured simply modulates the magnitude of some external power source. This is illustrated by examples.

An example of a passive instrument is the pressure- measuring device shown in figure 2.1. The pressure of the fluid is translated into a movement of a pointer against a scale. The energy expended in moving the pointer is derived entirely from the change in pressure measured: there are no other energy inputs to the system.

An example of an active instrument is a float-type petrol-tank level indicator as

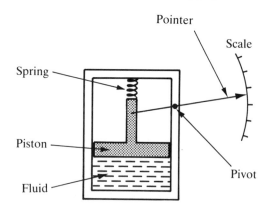

Figure 2.1 Passive pressure gauge

sketched in figure 2.2. Here, the change in petrol level moves a potentiometer arm, and the output signal consists of a proportion of the external voltage source applied across the two ends of the potentiometer. The energy in the output signal comes from the external power source: the primary transducer float system is merely modulating the value of the voltage from this external power source.

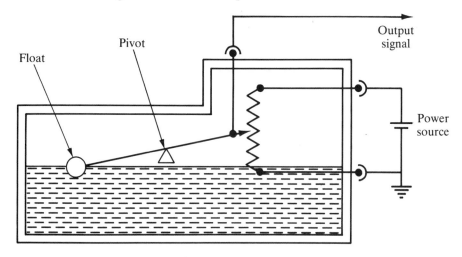

Figure 2.2 Petrol-tank level indicator

In active instruments, the external power source is usually in electrical form, but in some cases it can be other forms of energy such as a pneumatic or hydraulic one.

One very important difference between active and passive instruments is the level of measurement resolution which can be obtained. With the simple pressure gauge shown, the amount of movement made by the pointer for a particular pressure change is closely defined by the nature of the instrument. Whilst it is possible to increase measurement resolution by making the pointer longer, such that the pointer tip moves through a longer arc, the scope for such improvement is clearly limited by the practical limit of how long the pointer can conveniently be. In an active instrument, however, adjustment of the magnitude of the external energy input allows much greater control over measurement resolution. Whilst the scope for improving measurement resolution is much greater incidentally, it is not infinite because of limitations placed on the magnitude of the external energy input, in consideration of heating effects and for safety reasons.

In terms of cost, passive instruments are normally of a more simple construction than active ones and are therefore cheaper to manufacture. Choice between active and passive instruments for a particular application therefore involves carefully balancing the measurement-resolution requirements against cost.

Null/deflection type instruments

The pressure gauge just mentioned is a good example of a deflection type of instrument, where the value of the quantity being measured is displayed in terms of the amount of movement of a pointer.

An alternative type of pressure gauge is the dead-weight gauge shown in figure 2.3 which is a null-type instrument. Here, weights are put on top of the piston until the downward force balances the fluid pressure. Weights are added until the piston reaches a datum level, known as the null point. Pressure measurement is made in terms of the value of the weights needed to reach this null position.

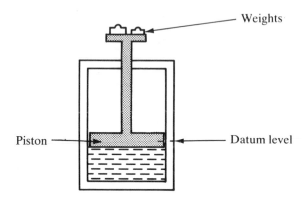

Figure 2.3   Dead-weight pressure gauge

The accuracy of these two instruments depends on different things. For the first one it depends on the linearity and calibration of the spring, whilst for the second it relies on the calibration of the weights. As calibration of weights is much easier than careful choice and calibration of a linear-characteristic spring, this means that the second type of instrument will normally be the more accurate. This is in accordance with the general rule that null-type instruments are more accurate than deflection types.

Monitoring/control instruments

An important distinction between different instruments is whether they are suitable only for monitoring functions or whether their output is in a form that can be directly included as part of an automatic control system. Instruments which only give an audio or visual indication of the magnitude of the physical quantity measured, such as a liquid-in-glass thermometer, are only suitable for monitoring purposes. This class normally includes all null-type instruments and most passive transducers.

For an instrument to be suitable for inclusion in an automatic control system, its output must be in a suitable form for direct input into the controller. Usually, this means that an instrument with an electrical output is required, although other forms of output such as optical or pneumatic signals are used in some systems.

Analog/digital instruments

An analog instrument gives an output which varies continuously as the quantity being measured changes. The output can have an infinite number of values within the range that the instrument is designed to measure. The deflection type of pressure gauge described earlier in this chapter is a good example of an analog instrument. As the input value changes, the pointer moves with a smooth continuous motion. Whilst the pointer can therefore be in an infinite number of positions within its range of movement, the number of different positions which the eye can discriminate between is strictly limited, this discrimination being dependent upon how large the scale is and how finely it is divided.

A digital instrument has an output which varies in discrete steps and so can only have a finite number of values. The rev-counter sketched in figure 2.4 is an example of a digital instrument. A cam is attached to the revolving body whose motion is being measured, and on each revolution the cam opens and closes a switch. The switching operations are counted by an electronic counter. This system can only count whole revolutions and cannot discriminate any motion which is less than a full revolution.

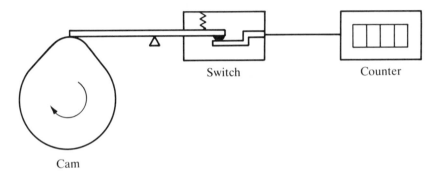

Figure 2.4   Rev-counter

The distinction between analog and digital instruments has become particularly important with the rapid growth in the application of microcomputers to automatic control systems. Any digital computer system, of which the microcomputer is but one example, performs its computations in digital form. An instrument whose output is in digital form is therefore particularly advantageous in such applications, as it can be interfaced directly to the control computer. Analog instruments must be interfaced to the microcomputer by an analog-to-digital (A/D) converter, which converts the analog output signal from the instrument into an equivalent digital quantity which can be read into the computer. This conversion has several disadvantages. First, the A/D converter adds a significant cost to the system. Second, a finite time is involved in the process of converting an analog signal to a digital quantity, and this time can be critical in the control of fast processes where the accuracy of control depends on the speed of the controlling computer. Degrading the speed of operation of the control computer by imposing a requirement for A/D conversion thus degrades the accuracy by which the process is controlled.

2.2   Standards

Referring back to our earlier comments in the last chapter about the development of measurement systems, we stated that the earliest units of measurement used were

Table 2.1   Definitions of standard units

Length	The *meter* (m) is the length of path traveled by light in an interval of 1/299 792 458 seconds
Mass	The *kilogram* (kg) is equal to the mass of a platinum–iridium cylinder kept in the International Bureau of Standards, Sevres, Paris
Time	The *second* (s) is the duration of $9.192\,631\,77 \times 10^9$ cycles of radiation from vaporized cesium-133 (an accuracy of 1 in 10^{12} or one second in 36 000 years)
Temperature	The temperature difference between absolute zero and the triple point of water is defined as 273.16 *kelvin* (K)
Current	One *ampere* (A) is the current flowing through two infinitely long parallel conductors of negligible cross-section placed one meter apart in vacuum and producing a force of 2×10^{-7} newtons per meter length of conductor
Amount of substance	The *mole* is the number of atoms in a specified mass of carbon-12

convenient parts of the human torso such as the foot and hand. Such a system allowed an approximate level of equivalence to be established about the relative value of quantities of different commodities. The length and breadth of constructional timber, for instance, could be measured in units of feet, and cloth could be similarly measured in units of feet-squared. This allowed a basis for determining the relative values of timber and cloth for barter trade purposes. However, such a system was clearly unfair when a person with large hands exchanged timber for cloth from a person with small hands (assuming that each used his own hands for measuring the commodity he was exchanging).

There was an urgent need to establish units of measurement which were based on non-varying quantities, and the first of these was a unit of length (the meter) defined as 10^{-7} times the polar quadrant of the Earth. A platinum bar made to this length was established as a standard of length in the early part of the nineteenth century. This was superseded by a superior quality standard bar in 1889, manu- factured from a platinum–iridium alloy. Since that time, technological research has enabled further improvements to be made in the standard used for defining length. First, in 1960, a standard meter was redefined in terms of $1.650\,763\,73 \times 10^6$ wavelengths of the radiation from krypton-86 in vacuum. More recently, in 1983, the meter was redefined yet again as the length of the path traveled by light in an interval of 1/299 792 458 seconds.

In a similar fashion, standard units for the measurement of other physical quantities have been defined and progressively improved over the years. The latest standards for defining the units used for measuring a range of physical variables are given in table 2.1.

Standard units

The early establishment of standards for the measurement of physical quantities proceeded in several countries at broadly parallel times, and, in consequence, several sets of units emerged for measuring the same physical variable. For instance, length could be measured in yards or meters or several other units. Apart from the major units of length, subdivisions of standard units exist such as feet, inches, centimeters and millimeters, with a fixed relationship between each fundamental unit and its subdivisions. Yards, feet and inches belong to the Imperial System of units, which is characterized by having varying and cumbersome multiplication factors relating fundamental units to subdivisions such as 1760 (miles to yards), 3 (yards to feet) and 12 (feet to inches). The Imperial standard of units was used predominantly in Britain until a few years ago, but since that time a strong effort has been made to standardize on the Système International d'Unités (International System of Units) which is normally abbreviated to SI units. This system includes units such as the meter, which is related to its centimeter and millimeter subdivisions by mathematically convenient powers of 10. The SI system of units is the one now adopted by international agreement and will be used exclusively in this book. Conversion tables relating common Imperial units to equivalent SI units are to be found in appendix 1.

Apart from the establishment of a common set of measurement units (the SI system), international agreements have also established several other common conventions. One such agreement is the convention used to mark component values and tolerances on electric circuit components such as resistors and capacitors, as detailed in appendix 2. Further examples are internationally accepted codes (G-codes) for programming numerically controlled machine tools, and a system of standard symbols for representing components in electric circuit drawings.

2.3   Static characteristics of instruments

If we have a thermometer in a room and its reading shows a temperature of 20°C, then it does not really matter whether the true temperature of the room is 19.5°C or 20.5°C. Such small variations around 20°C are too small to affect whether we feel warm enough or not. Our bodies cannot discriminate between such close levels of temperature and therefore a thermometer with an accuracy of ±0.5°C is perfectly adequate. If we had to measure the temperature of certain chemical processes, however, a variation of 0.5°C might have a significant effect on the rate of reaction or even the products of a process. A measurement accuracy much better than ±0.5°C is therefore clearly required.

Accuracy of measurement is thus one consideration in the choice of instrument for a particular application. Other quantities such as sensitivity, linearity and reaction to ambient temperature changes are further considerations. These attributes are collectively known as the static characteristics of instruments, and are defined as follows.

Accuracy

Accuracy is the extent to which a reading might be wrong, and is often quoted as a percentage of the full-scale reading of an instrument.

If, for example, a pressure gauge of range 0–10 bar has a quoted inaccuracy of

±1.0% f.s. (±1% of full-scale reading), then the maximum error to be expected in *any* reading is 0.1 bar. This means that when the instrument is reading 1.0 bar, the possible error is 10% of this value. For this reason, it is an important system design rule that instruments are chosen such that their range is appropriate to the spread of values being measured, in order that the best possible accuracy be maintained in instrument readings. Thus, if we were measuring pressures with expected values between 0 and 1 bar, we would not use an instrument with a range of 0–10 bar.

Tolerance

Tolerance is a term which is closely related to accuracy and defines the maximum error which is to be expected in some value. Whilst it is not, strictly speaking, a static characteristic of measuring instruments, it is mentioned here because the accuracy of some instruments is sometimes quoted as a tolerance figure.

  Tolerance, when used correctly, describes the maximum deviation of a manufactured component from some specified value. Crankshafts, for instance, are machined with a diameter tolerance quoted as so many microns (micrometers), and electric circuit components such as resistors have tolerances of perhaps 5%. One resistor chosen at random from a batch having a nominal value 1000 Ω and tolerance 5% might have an actual value anywhere between 950 Ω and 1050 Ω.

Range or span

The range or span of an instrument defines the minimum and maximum values of a quantity that the instrument is designed to measure.

Bias

Bias describes a constant error which exists over the full range of measurement of an instrument. This error is normally removable by calibration.

  Bathroom scales are a common example of instruments which are prone to bias. It is quite usual to find that there is a reading of perhaps 1 kg with no one stood on the scales. If someone of known weight 70 kg were to get on the scales, the reading would be 71 kg, and if someone of known weight 100 kg were to get on the scales, the reading would be 101 kg. This constant bias of 1 kg can be removed by calibration: in the case of bathroom scales this normally means turning a thumbwheel, with the scales unloaded, until the reading is zero.

Linearity

It is normally desirable that the output reading of an instrument is linearly proportional to the quantity being measured. The '×'s marked on figure 2.5 show a plot of the typical output readings of an instrument when a sequence of input quantities are applied to it. Normal procedure is to draw a good fit straight line through the '×'s, as shown in figure 2.5. (Whilst this can often be done with reasonable accuracy by eye, it is always preferable to apply a mathematical least-squares line-fitting technique, as described in chapter 7.) The non-linearity is then defined as the

Output
reading

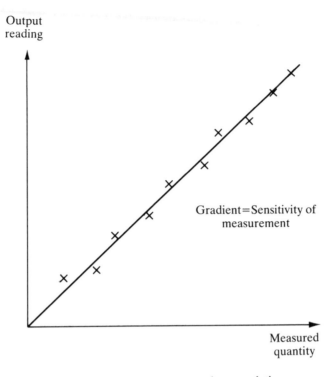

Gradient=Sensitivity of
measurement

Measured
quantity

Figure 2.5   Instrument output characteristic

maximum deviation of any of the output readings marked × from this straight line. Non-linearity is usually expressed as a percentage of full-scale reading. Referring back to the definition of accuracy, it can be seen that accuracy and linearity are closely related quantities.

Sensitivity of measurement

The sensitivity of measurement is a measure of the change in instrument output which occurs when the quantity being measured changes by a given amount. Sensitivity is thus the ratio:

$$\frac{\text{Scale deflection}}{\text{Value of measurand causing deflection}}$$

The sensitivity of measurement is therefore the slope of the straight line drawn on figure 2.5. If, for example, a pressure of 2 bar produces a deflection of 10 degrees in a pressure transducer, the sensitivity of the instrument is 5 degrees/bar.

Sensitivity to disturbance

All calibrations and specifications of an instrument are only valid under controlled conditions of temperature, pressure, etc. These standard ambient conditions are usually defined in the instrument specification. As variations occur in the ambient temperature etc., certain static instrument characteristics change, and the sensitivity to disturbance is a measure of the magnitude of this change.

Such environmental changes affect instruments in two main ways, known as zero drift and sensitivity drift.

Zero drift

Zero drift describes the effect where the zero reading of an instrument is modified by a change in ambient conditions.

Typical units by which zero drift is measured are volts/°C, in the case of a voltmeter affected by ambient temperature changes. The effect of zero drift is to impose a bias in the instrument output readings: this is normally removable by recalibration in the usual way. A typical change in the output characteristic of a pressure gauge subject to zero drift is shown in figure 2.6(a).

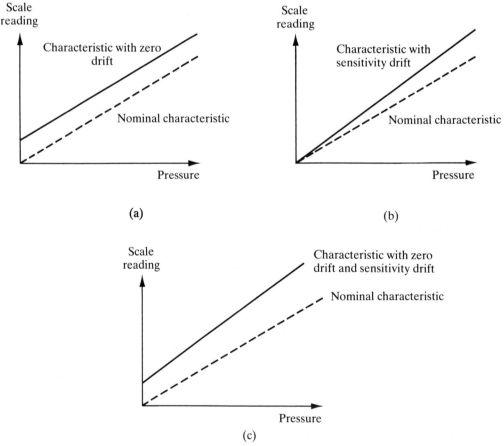

Figure 2.6   (a) Zero drift (b)   Sensitivity drift
(c)   Zero drift plus sensitivity drift

Sensitivity drift (scale factor drift)

Sensitivity drift defines the amount by which an instrument's sensitivity of measurement varies as ambient conditions change. An alternate name for this phenomenon is scale factor drift.

Many components within an instrument are affected by environmental fluctuations, such as temperature changes: for instance, the modulus of elasticity of a spring is temperature-dependent. Figure 2.6(b) shows what effect sensitivity drift can have on the output characteristic of an instrument.

Sensitivity drift is measured in units of the form (degrees/bar)/°C.

If an instrument suffers both zero drift and sensitivity drift at the same time, then the typical modification of the output characteristic is shown in figure 2.6(c).

Hysteresis

Figure 2.7 illustrates the output characteristic of an instrument which exhibits hysteresis. If the input measured quantity to the instrument is steadily increased from a negative value, the output reading varies in the manner shown in curve A. If the input variable is then steadily decreased, the output varies in the manner shown in curve B. The non-coincidence between these loading and unloading curves is known as hysteresis.

Two quantities, maximum input hysteresis and maximum output hysteresis, are defined as shown in figure 2.7. These are normally expressed as a percentage of the full-scale input or output reading, respectively.

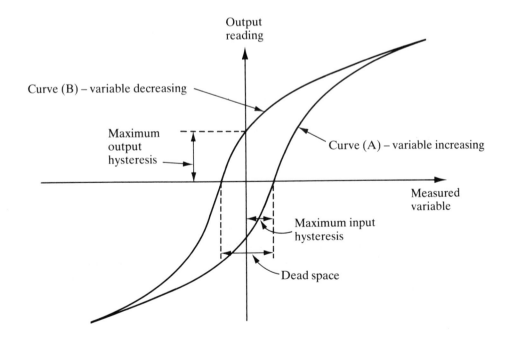

Figure 2.7   Instrument characteristic with hysteresis

Dead space

Dead space is defined as the range of different input values over which there is no change in output value.

Any instrument which exhibits hysteresis also displays dead space, as marked on figure 2.7. Some instruments which do not suffer from any significant hysteresis can still exhibit a dead space in their output characteristics, however. Backlash in gears is a typical cause of dead space, and results in the sort of instrument output characteristic shown in figure 2.8.

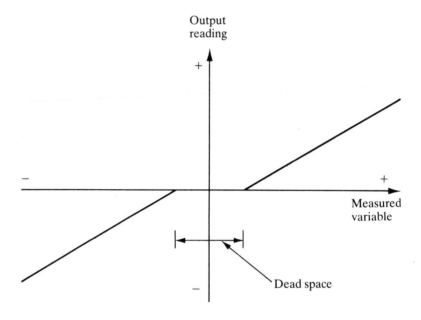

Figure 2.8   Instrument characteristic with dead space

Threshold

If the input to an instrument is gradually increased from zero, the input will have to reach a certain minimum level before the change in the instrument output reading is of a large enough magnitude to be detectable. This minimum level of input is known as the threshold of the instrument. Manufacturers vary in the way that they specify threshold for instruments. Some quote absolute values, whereas others quote threshold as a percentage of full-scale readings.

As an illustration, a car speedometer typically has a threshold of about 15 km/h. This means that, if the vehicle starts from rest and accelerates, no output reading is observed on the speedometer until the speed reaches 15 km/h.

Resolution

When an instrument is showing a particular output reading, there is a lower limit on the magnitude of the change in the input measured quantity which produces an observable change in the instrument output. Like threshold, resolution is sometimes specified as an absolute value and sometimes as a percentage of f.s. deflection.

One of the major factors influencing the resolution of an instrument is how finely its output scale is divided into subdivisions. Using a car speedometer as an example again, this has subdivisions of typically 20 km/h. This means that when the needle is between the scale markings, we cannot estimate speed more accurately than to the nearest 5 km/h. This figure of 5 km/h thus represents the resolution of the instrument.

Cost, durability and maintenance

The static characteristics discussed so far are those features which form the technical basis for a comparison between the relative merits of different instruments. In assessing the relative suitability of different instruments for a particular measurement situation, however, considerations of cost, durability and maintenance are also of great importance, and they are therefore lumped into this section on instrument static characteristics even though they are not technically instrument characteristics.

Cost is very strongly correlated with the performance of an instrument, as measured by its static characteristics. Increasing the accuracy or resolution of an instrument, for example, can only be done at a penalty of increasing its manufacturing cost. Instrument choice therefore proceeds by specifying the minimum characteristics required by a measurement situation and then searching manufacturers' catalogs to find an instrument whose characteristics match those required. To select an instrument with characteristics superior to those required would only mean paying more than necessary for a level of performance greater than that needed.

As well as purchase cost, other important factors in the assessment exercise are instrument durability and the maintenance requirements. Assuming that one had $10 000 to spend, one would not spend $8000 on a new motor car whose projected life was five years if a car of equivalent specification with a projected life of 10 years was available for $10 000. Likewise, durability is an important consideration in the choice of instruments. The projected life of instruments often depends on the conditions in which the instrument will have to operate. Maintenance requirements must also be taken into account, as they also have cost implications.

As a general rule, a good assessment criterion is obtained if the total purchase cost and estimated maintenance costs of an instrument over its life are divided by the period of its expected life. The figure obtained is thus a cost per year. However, this rule becomes modified where instruments are being installed on a process whose life is expected to be limited, perhaps in the manufacture of a particular model of car. Then, the total costs can only be divided by the period of time an instrument is expected to be used, unless an alternative use for the instrument is envisaged at the end of this period.

To summarize, therefore, instrument choice is a compromise between performance characteristics, ruggedness and durability, maintenance requirements and purchase cost. To carry out such an evaluation properly, the instrument engineer must have a wide knowledge of the range of instruments available for measuring particular physical quantities, and he/she must also have a deep understanding of how instrument characteristics are affected by particular measurement situations and operating conditions.

2.4 Dynamic characteristics of instruments

The static characteristics of measuring instruments are concerned only with the steady-state reading that the instrument settles down to, such as the accuracy of the reading, etc.

The dynamic characteristics of a measuring instrument describe its behavior between the time a measured quantity changes value and the time when the instrument output attains a steady value in response.

In any linear, time-invariant measuring system, the following general relation can be written between input and output for time $t > 0$:

$$a_n \frac{d^n q_0}{dt^n} + a_{n-1} \frac{d^{n-1} q_0}{dt^{n-1}} + \ldots + a_1 \frac{dq_0}{dt} + a_0 q_0 = b_m \frac{d^m q_i}{dt^m} + b_{m-1} \frac{d^{m-1} q_i}{dt^{m-1}} + \ldots + b_1 \frac{dq_i}{dt} + b_0 q_i \quad (2.1)$$

where q_i is the measured quantity, q_0 is the output reading and $a_0, \ldots, a_n, b_0, \ldots, b_m$ are constants.

The reader whose mathematical background is such that the above equation appears daunting should not worry unduly, as only certain special, simplified cases of it are applicable in normal measurement situations. The major point of importance is to have a practical appreciation of the manner in which various different types of instrument respond when the measurand applied to them varies.

If we limit consideration to that of step changes in the measured quantity only, then equation (2.1) reduces to:

$$a_n \frac{d^n q_0}{dt^n} + a_{n-1} \frac{d^{n-1} q_0}{dt^{n-1}} + \ldots + a_1 \frac{dq_0}{dt} + a_0 q_0 = b_0 q_i \quad (2.2)$$

Further simplification can be made by taking certain special cases of equation (2.2) which collectively apply to nearly all measurement systems.

Zero-order instrument

If all the coefficients $a_1, \ldots, a_n$ other than a_0 in equation (2.2) are assumed zero, then:

$$a_0 q_0 = b_0 q_i \quad \text{or} \quad q_0 = \frac{b_0 q_i}{a_0} = K q_i \quad (2.3)$$

where K is a constant known as the instrument sensitivity as defined earlier.

Any instrument which behaves according to equation (2.3) is said to be of zero-order type. A potentiometer, which measures motion, is a good example of such an instrument, where the output voltage changes instantaneously as the slider is displaced along the potentiometer track.

First-order instrument

If all the coefficients $a_2, \ldots, a_n$ except for a_0 and a_1 are assumed zero in equation (2.2) then:

$$a_1 \frac{dq_0}{dt} + a_0 q_0 = b_0 q_i \qquad (2.4)$$

Any instrument which behaves according to equation (2.4) is known as a first-order instrument.

If d/dt is replaced by the operator D in equation (2.4), we get:

$$a_1 D q_0 + a_0 q_0 = b_0 q_i$$

and rearranging:

$$q_0 = \frac{(b_0/a_0)q_i}{1 + (a_1/a_0)D} \qquad (2.5)$$

Defining $K = b_0/a_0$ as the static sensitivity and $\tau = a_1/a_0$ as the time constant of the system, equation (2.5) becomes:

$$q_0 = \frac{Kq_i}{1 + \tau D} \qquad (2.6)$$

If equation (2.6) is solved analytically, the output quantity q_0 in response to a step change in q_i varies with time in the manner shown in figure 2.9. The time constant τ of the step response is the time taken for the output quantity q_0 to reach 63% of its final value.

The thermocouple (see chapter 15) is a good example of a first-order instrument. It is well known that, if a thermocouple at room temperature is plunged into boiling water, the output e.m.f. does not rise instantaneously to a level indicating 100°C but instead approaches a reading indicating 100°C in a manner similar to that shown in figure 2.9.

A large number of other instruments also belong to this first-order class: this is of particular importance in control systems where it is necessary to take account of the time lag that occurs between a measured quantity changing in value and the measuring instrument indicating the change. Fortunately, the time constant of many first-order instruments is small relative to the dynamics of the process being measured, and so no serious problems are created.

Example

A balloon is equipped with temperature- and altitude- measuring instruments and has radio equipment which can transmit the output readings of these instruments back to ground. The balloon is initially anchored to the ground with the instrument output readings in steady state. The altitude- measuring instrument is approximately zero-order and the temperature transducer first-order with a

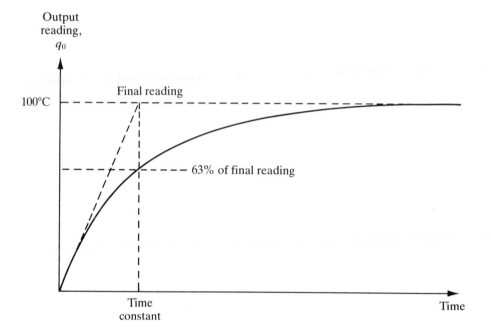

Figure 2.9   First-order instrument characteristic

time constant of 15 s. The temperature on the ground, T_0, is 10°C and the temperature T_x at an altitude of x meters is given by the relation:

$$T_x = T_0 - 0.01x$$

(a) If the balloon is released at time zero, and thereafter rises upwards at a velocity of 5 m/s, draw a table showing the temperature and altitude measurements reported at intervals of 10 s over the first 50 s of travel. Show also in the table the error in each temperature reading.
(b) What temperature does the balloon report at an altitude of 5000 m?

Solution
In order to answer this question, it is assumed that the solution of a first-order differential equation has been presented to the reader in a mathematics course. If the reader is not so equipped, the following solution will be difficult to follow.

Let the temperature reported by the balloon at some general time t be T_r. Then T_x is related to T_r by the relation:

$$T_r = \frac{T_x}{1+\tau D} = \frac{T_0 - 0.01x}{1+\tau D} = \frac{10 - 0.01x}{1 + 15D}$$

It is given that $x = 5t$. Thus:

$$T_r = \frac{10 - 0.05t}{1 + 15D}$$

The transient or complementary-function part of the solution ($T_x = 0$) is given by:

$$T_{r,cf} = Ce^{-t/15}$$

The particular-integral part of the solution is given by:

$$T_{r,pi} = 10 - 0.05 (t - 15)$$

Thus the whole solution is given by:

$$T_r = T_{r,cf} + T_{r,pi} = Ce^{-t/15} + 10 - 0.05 (t - 15)$$

Applying initial conditions: at $t = 0$, $T_r = 10$, i.e.

$$10 = Ce^{-0} + 10 - 0.05 (-15)$$

Thus $C = -0.75$ and the solution can be written as:

$$T_r = 10 - 0.75e^{-t/15} - 0.05 (t - 15)$$

(a) Using the above expression to calculate T_r for various values of t, the following table can be constructed:

Time (s)	Altitude (m)	Temp. reading (°C)	Temp. error (°C)
0	0	10	0
10	50	9.86	0.36
20	100	9.55	0.55
30	150	9.15	0.65
40	200	8.70	0.70
50	250	8.22	0.72

(b) At 5000 m, $t = 1000$ s. Calculating T_r from the above expression:

$$T_r = 10 - 0.75e^{-1000/15} - 0.05(1000 - 15)$$

The term in e approximates to zero and so T_r can be written as:

$$T_r \simeq 10 - 0.05(985) = -39.25°C$$

This result might have been inferred from the table above where it can be seen that the error is converging toward a value of 0.75. For large values of t, the transducer reading lags the true temperature value by a period of time equal to the time constant of 15 s. In this time, the balloon travels a distance of 75 m and the temperature falls by 0.75°C. Thus for large values of t, the output reading is always 0.75°C less than it should be.

Second-order instrument

If all coefficients $a_3, \ldots, a_n$ other than a_0, a_1 and a_2 in equation (2.2) are assumed zero, then we get:

$$a_2 \frac{d^2 q_0}{dt^2} + a_1 \frac{dq_0}{dt} + a_0 q_0 = b_0 q_i \tag{2.7}$$

Applying the D operator again:

$$a_2 D^2 q_0 + a_1 D q_0 + a_0 q_0 = b_0 q_i$$

and rearranging:

$$q_0 = \frac{b_0 q_i}{a_0 + a_1 D + a_2 D^2} \tag{2.8}$$

It is convenient to re-express the variables a_0, a_1, a_2 and b_0 in equation (2.8) in terms of three parameters K (static sensitivity), ω (undamped natural frequency) and ϵ (damping ratio) where

$$K = b_0/a_0$$

$$\omega = a_0/a_2$$

$$\epsilon = \frac{a_1}{2 a_0 a_2}$$

Re-expressing equation (2.8) in terms of K, ω and ϵ we get:

$$\frac{q_0}{q_i} = \frac{K}{D^2/\omega^2 + 2\epsilon D/\omega + 1} \tag{2.9}$$

This is the standard equation for a second-order system and any instrument whose response can be described by it is known as a second-order instrument.

If equation (2.9) is solved analytically, the shape of the step response obtained depends on the value of the damping ratio parameter ϵ. The output responses of a second order instrument for various values of ϵ are shown in figure 2.10. For case A where $\epsilon = 0$, there is no damping and the instrument output exhibits constant-amplitude oscillations when disturbed by any change in the physical quantity measured. For light damping of $\epsilon = 0.2$, represented by case B, the response to a step change in input is still oscillatory but the oscillations gradually die down. Further increase in the value of ϵ reduced oscillations and overshoot still more, as shown by curves C and D, and finally the response becomes very overdamped as shown by curve E where the output reading creeps up slowly toward the correct reading. Clearly, the extreme response curves A and E are grossly unsuitable for any measuring instrument. If an instrument were to be only ever subjected to step inputs, then

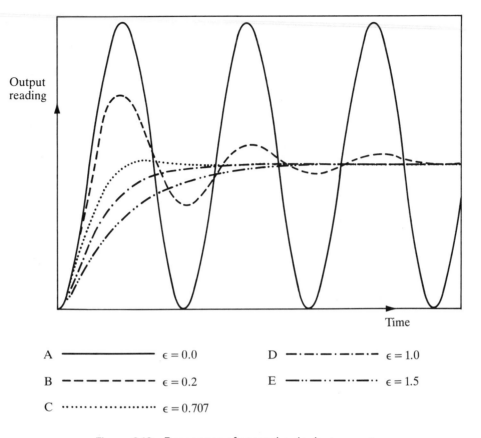

Figure 2.10   Responses of second order instruments

the design strategy would be to aim toward a damping ratio of 0.707, which gives the critically damped response C. Unfortunately, most of the physical quantities which instruments are required to measure do not change in the mathematically convenient form of steps, but rather in the form of ramps of varying slopes. As the form of the input variable changes, so the best value for ϵ varies, and choice of ϵ becomes one of compromise between those values that are best for each type of input variable behavior anticipated. Commercial second-order instruments, of which the accelerometer is a common example, are generally designed to have a damping ratio ϵ somewhere in the range 0.6–0.8.

2.5   Calibration

The foregoing discussion has described the static and dynamic characteristics of measuring instruments in some detail. An important qualification which has been omitted from this discussion, however, is the statement that an instrument only conforms to stated static and dynamic patterns of behavior after it has been calibrated.

It can normally be assumed that a new instrument will have been calibrated when it is obtained from an instrument manufacturer, and will therefore initially behave according to the characteristics stated in the specifications. During use, however, its behavior will gradually diverge from the stated specification for a variety of reasons. Such reasons include mechanical wear, and the effects of dirt, dust, fumes and chemicals in the operating environment. The rate of divergence from standard specifications varies according to the type of instrument, the frequency of usage and the severity of the operating conditions. However, there will come a time, determined by practical knowledge, when the characteristics of the instrument will have drifted from the standard specification by an unacceptable amount. When this situation is reached, it is necessary to recalibrate the instrument back to the standard specifications. Such recalibration is performed by adjusting the instrument at each point in its output range until its output readings are the same as those of a second standard instrument to which the same inputs are applied. This second instrument is one kept solely for calibration purposes whose specifications are accurately known. The calibration story does not end there, however. Whilst this second instrument can be used as an accurate standard in the short term, it too can suffer drift in the long term and therefore must itself be calibrated at less frequent intervals against a yet more accurate standard instrument. In fact, the process of instrument calibration follows a hierarchical structure, as shown in figure 2.11, where each instrument in the chain is more accurate and better calibrated than the one below.

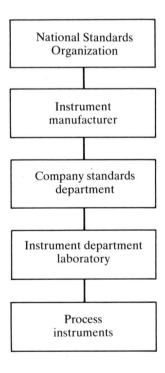

Figure 2.11   Instrument calibration chain

At the lowest point in this chain are the instruments which are installed on process plant to measure process variables. The instrument department responsible for the maintenance of such process instruments will normally maintain a standards laboratory containing standard instruments which are kept solely for calibration purposes, and periodically all process instruments will be temporarily removed from the process and recalibrated in this laboratory. These calibration instruments will themselves be periodically checked against instruments maintained to an even higher standard of specification in a calibration service provided by instrument manufacturers. Where the company concerned is a large one, with several geographically distributed instrument departments, an intermediate level of calibration will be provided by a company standards department. Finally, at the top of the tree, most countries have a National Standards Organization (e.g. National Bureau of Standards in U.S.A., National Physical Laboratory in U.K.). Such organizations maintain laboratories which contain instruments conforming to specifications to the highest accuracy attainable and provide a calibration service where manufacturers' 'standard' instruments are checked.

In such calibration of instruments, the environmental conditions under which the calibration is performed must be carefully controlled. Study of any instrument specification as supplied by a manufacturer defining the characteristics of an instrument will reveal that the characteristics are specified for one particular set of operating conditions only. As discussed in more detail in chapter 6, the characteristics of any instrument change to a certain extent as the operating conditions of temperature, pressure and humidity etc. change. Calibration is only valid therefore if it is performed under the environmental conditions stated in the manufacturer's specification for the instrument.

Exercises

2.1 Explain what is meant by:
(a) active instruments
(b) passive instruments.
Give examples of each and discuss the relative merits of these two classes of instruments.

2.2 Discuss the advantages and disadvantages of null and deflection types of measuring instrument.

2.3 Briefly define and explain 10 static characteristics of measuring instruments.

2.4 A tungsten/5% rhenium–tungsten/26% rhenium thermocouple has an output e.m.f. as shown in the following table when its hot (measuring) junction is at the temperatures shown. Determine the sensitivity of measurement for the thermocouple in mV/°C.

E.m.f (mV)	4.37	8.74	13.11	17.48
Temp. (°C)	250	500	750	1000

2.5 An unmanned submarine is equipped with temperature- and depth-measuring instruments and has radio equipment which can transmit the output readings of these instruments back to the surface. The submarine is initially floating on the surface of the sea with the instrument output readings in steady state. The depth-measuring instrument is approximately zero-order and the temperature transducer first-order with a time constant of 50 s. The water temperature on the sea surface, T_0, is 20°C and the temperature T_x at a depth of x meters is given by the relation:

$$T_x = T_0 - 0.01x$$

(a) If the submarine starts diving at time zero, and thereafter goes down at a velocity of 0.5 m/s, draw a table showing the temperature and depth measurements reported at intervals of 100 s over the first 500 s of travel. Show also in the table the error in each temperature reading.

(b) What temperature does the submarine report at a depth of 1000 m?

2.6 Write down the general differential equation describing the dynamic response of a second-order measuring instrument and state the expressions relating the static sensitivity, undamped natural frequency and damping ratio to the parameters in this differential equation. Sketch the instrument response for the cases of heavy damping, critical damping and light damping, and state which of these is the usual target when a second-order instrument is being designed.

3

Intelligent Instruments

The impact that the microprocessor has had on the domestic consumer is very evident, with their inclusion in washing machines, vehicle fuel-control systems and home computers to name but a few applications. The advent of the microprocessor has had an equally significant, though perhaps not so obviously apparent, impact on the field of instrumentation. It is therefore fitting that a chapter of this book should be devoted to describing the principles of operation of microprocessors and explaining how they are included and programmed as a microcomputer within a measuring instrument, thereby producing what has come to be known as an intelligent instrument.

An intelligent instrument comprises all the usual elements of a measurement system as shown in figure 1.3, and is only distinguished from dumb (non-intelligent) measurement systems by the inclusion of a microprocessor to fulfil the signal processing function. The effect of this computerization of the signal processing function is an improvement in the quality of the instrument output measurements and a general simplification of the signal processing task. Some examples of the signal processing which a microprocessor can readily perform include correction of the instrument output for bias caused by temperature variations, and conversion to produce a linear output from a transducer whose characteristic is fundamentally non-linear. A fuller discussion about the techniques of digital signal processing is to be found in section 5.6.

3.1 Elements of a microcomputer

The primary function of a digital computer is the manipulation of data and the three elements which are essential to the fulfillment of this task are the central processing unit, the memory and the input–output interface, as shown in figure 3.1. These elements are collectively known as the computer hardware and each element exists physically as one or more integrated circuit chips mounted on a printed circuit board. Where the central processing unit (CPU) consists of a single microprocessor, it is usual to regard the system as a microcomputer. The distinction between microcomputer, minicomputer and mainframe computer is a very arbitrary division made according to relative computer power. However, this classification has become somewhat meaningless, with present-day 'microcomputers' being more powerful than mainframe computers of only a few years ago.

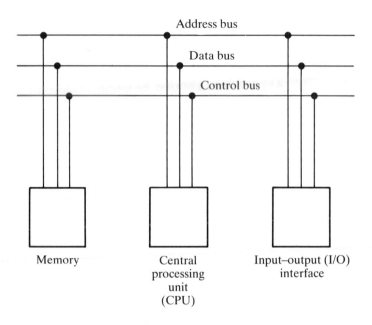

Figure 3.1   Elements of a microcomputer

The *central processing unit* (CPU) part of a computer can be regarded as the brain of the system. The CPU determines what computational operations are carried out and the sequence in which the operations are executed. During such operation, the CPU makes use of one or more special storage locations within itself known as *registers*. Another part of the CPU is the *arithmetic and logic unit* (ALU) which is where all arithmetic operations are evaluated. The CPU operates according to a sequential list of required operations defined by a computer program, known as the computer software. This program is held in the second of the three system components known as the computer memory.

The *computer memory* also serves several other functions besides this role of holding the computer program. One of these is to provide temporary storage locations which the CPU uses to store variables during execution of the computer program. A further common use of memory is to store data tables which are used for scaling and variable conversion purposes during program execution.

Memory can be visualized as a consecutive sequence of boxes in which various items are stored, as shown in figure 3.2 for a typical memory size of 65 536 storage units. If this storage mechanism is to be useful, then it is essential that a means is provided for giving a unique label to each storage box. This is achieved by labeling the first box as 0, the next one as 1 and so on for the rest of the storage locations. These numbers are known as the *memory addresses*. Whilst these can be labeled by decimal numbers, it is more usual to use hexadecimal notation (see section 3.2).

Two main types of computer memory exist and there are important differences between these. The two kinds are *random-access memory* (RAM) and *read-only memory* (ROM). The CPU can both read from and write to the former, but it can only read from the latter. The importance of ROM becomes apparent if the behavior

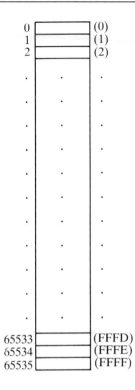

Figure 3.2   Schematic representation of computer memory
(numbers in parentheses are memory addresses in hexadecimal notation)

of each kind of memory when the power supply is turned off is considered. At power-off time, RAM loses its contents but ROM maintains them, and this is the value of ROM. Intelligent instruments normally use ROM for storage of the program and data tables and just have a small amount of RAM which is used by the CPU for temporary variable storage during program execution.

The third essential element of a computer system is the *input–output (I/O) interface*, which allows the computer to communicate with the outside world by reading in data values and outputting results after the appropriate computation has been executed. In the case of a microcomputer performing a signal processing function within an intelligent instrument, this means reading in the values obtained from one or more transducers and outputting a processed value for presentation at the instrument output. All such external peripherals are identified by a unique number as for memory addresses.

Communication between these three computer elements is provided by three electronic highways known as the data bus, the address bus and the control bus. At each data transfer operation executed by the CPU, two items of information must be conveyed along the electronic highway, the item of data being transferred and the address where it is being sent. Whilst both of these items of information could be conveyed along a single bus, it is more usual to use two buses which are called the data bus and the address bus. The timing of data transfer operations is important, particularly when transfers take place to peripherals such as disk drives and keyboards where the CPU often has to wait until the peripheral is free before it can initialize a data transfer. This timing information is carried by a third highway known as the control bus.

The latest trend made possible by advances in very-large-scale integration (VLSI) technology is to incorporate all three functions of central processor unit, memory and I/O within a single chip (known as the computer on a chip). These are already well established in domestic appliances and vehicle fuel-control systems and their exploitation within intelligent instruments is likely to grow rapidly.

3.2   Microcomputer operation

As has already been mentioned, the fundamental role of a computer is the manipulation of data. Numbers are used both in quantifying items of data and also in the form of codes which define the computational operations which are to be executed. All numbers which are used for these two purposes must be stored within the computer memory and also transported along the communication buses. A detailed consideration of the conventions used for representing numbers within the computer is therefore required.

Number systems

The decimal system is the best-known number system, but it is not very suitable for use by digital computers. It uses a base of 10, such that each digit in a number can have any one of 10 values within the range 0–9. Items of electronic equipment such as the digital counter, which are often used as computer peripherals, have liquid-crystal display elements which can each display any of the 10 decimal digits, and therefore a four-element display can directly represent decimal numbers in the range 0–9999. The decimal system is therefore perfectly suitable for use with such output devices.

The fundamental unit of data storage within a digital computer is a memory element known as a *bit*. This holds information by switching between one of two possible states. Each storage unit can therefore only represent two possible values and all data to be entered into memory must be organized into a format which recognizes this restriction. This means that numbers must be entered in binary format, where each digit in the number can only have one of two values, 0 or 1. Whilst conversion of numbers between decimal and binary representations is tedious for humans, the binary representation is particularly convenient for computers because bits can be represented very simply electronically as either zero or non-zero voltages.

For data storage purposes, memory elements are combined into larger units known as *bytes*, which are usually considered to consist of eight bits each. Each bit holds one binary digit, and therefore a memory unit consisting of eight bits can store eight-digit binary numbers in the range of 00000000 to 11111111 (equivalent to decimal numbers in the range of 0 to 255). A binary number in this system of 10010011 for instance would correspond with the decimal number 147. This range is clearly inadequate for most purposes, including measurement systems, because even if all data could be conveniently scaled the maximum resolution obtainable is only one part in 128. Numbers are therefore normally stored in units of either two or four bytes, which allow the storage of integer (whole) numbers in the range of 0–65 535 or 0–4294 967 296.

No means have been suggested so far for expressing the sign of numbers, which is clearly necessary in the real world where negative as well as positive numbers occur. A simple way to do this is to reserve the most significant (left-hand bit) in a storage unit to define the sign of a number, with a '0' representing a positive number and a '1' a negative number. This alters the ranges of numbers representable in a one-byte storage unit to −127 to +127, as only seven bits are left to express the magnitude of the number, and also means that there are two representations of the value 0. In this system the binary number 10010011 translates to the decimal number −19 and 00010011 translates to +19. For reasons dictated by the mode of operation of the CPU, however, most computers use an alternative representation known as the two's complement form.

The *two's complement* of a number is most easily formed by going via an intermediate stage of the one's complement. The one's complement of a number is formed by reversing all digits in the binary representation of the magnitude of a number, changing '1's to '0's and '0's to '1's, and then changing the left-hand bit to a '1' if the original number was negative. The two's complement is then formed by adding 1 at the least significant (right-hand) end of the one's complement. As before for a one-byte storage unit, only seven bits are available for representing the magnitude of a number, but, because there is now only one representation of a zero, the decimal range representable is −128 to +127.

Examples
Find the two's complement eight-bit binary representations of the following decimal numbers:

$$56 \quad -56 \quad 73 \quad 119 \quad 27 \quad -47$$

Method of solution
Take first the decimal value 56.

Form seven-bit binary representation: 0111000

Reverse digits in this: 1000111

Add sign bit to left-hand end to form one's complement: 01000111

Form two's complement by adding 1 to one's complement:
$$01000111 + 1 = 01001000$$

Take next the decimal value −56.

Form seven-bit binary representation: 0111000

Reverse digits in this: 1000111

Add sign bit to left-hand end to form one's complement: 11000111

Form two's complement by adding 1 to one's complement:
$$11000111 + 1 = 11001000$$

Summary of solution for all values:

Decimal no.	Binary representation of magnitude (seven bit)	Digits reversed in seven-bit representation	One's complement (eight bit)	Two's complement (eight bit)
56	0111000	1000111	01000111	01001000
–56	0111000	1000111	11000111	11001000
73	1001001	0110110	00110110	00110111
119	1110111	0001000	00001000	00001001
27	0011011	1100100	01100100	01100101
–47	0101111	1010000	11010000	11010001

We have therefore established the binary code in which the computer stores positive and negative integers (whole numbers). However, it is frequently necessary also to handle real numbers (those with fractional parts). These are most commonly stored using the floating-point representation.

The *floating-point representation* divides each memory storage unit (notionally, not physically) into three fields, known as the sign field, the exponent field and the mantissa field. The sign field is always one-bit wide but there is no formal definition for the relative sizes of the other fields. However, a common subdivision of a 32-bit (four-byte) storage unit is to have a seven-bit exponent field and a 24-bit mantissa field, as shown below.

Sign   Exponent   Mantissa

1   7   24

The value contained in the storage unit is evaluated by multiplying the number in the mantissa field by 2 raised to the power of the number in the exponent field. Negative as well as positive exponents are obtained by biasing the exponent field by 64 (for a seven-bit field), such that a value of 64 is interpreted as an exponent of 0, a value of 65 as an exponent of 1, a value of 63 as an exponent of -1, etc. Suppose therefore that the sign bit field has a zero, the exponent field has a value of 0111110 (decimal 62) and the mantissa field has a value of 000000000000000001110111 (decimal 119), i.e. the contents of the storage unit are 00111110000000000000000001110111. The number stored is $+119 \times 2^{-2}$. Changing the first (sign) bit to a 1 would change the number stored to -119×2^{-2}.

If a human being were asked to enter numbers in these binary forms, however, the procedure would be both highly tedious and also very prone to error, and, in consequence, simpler ways of entering binary numbers have been developed. Two such ways are to use octal and hexadecimal numbers, which are translated to binary numbers at the input–output interface to the computer.

Octal numbers use a base of eight and consist of decimal digits in the range of 0–7 which each represent three binary digits. Thus a 24-bit binary number is represented by eight octal digits.

Hexadecimal numbers have a base of 16. They use decimal digits in the range 0–9 and letters in the range A–F which each represent four binary digits. The decimal digits 0–9 translate directly to the decimal values 0–9 and the letters A–F translate respectively to the decimal values 10–15. A 24-bit binary number requires six hexadecimal digits to represent it. The following table shows the octal, hexadecimal and binary equivalents of decimal numbers in the range of 0–15.

Decimal	Octal	Hexadecimal	Binary
0	0	0	0
1	1	1	1
2	2	2	10
3	3	3	11
4	4	4	100
5	5	5	101
6	6	6	110
7	7	7	111
8	10	8	1000
9	11	9	1001
10	12	A	1010
11	13	B	1011
12	14	C	1100
13	15	D	1101
14	16	E	1110
15	17	F	1111

Suppose that the 24-bit binary number 011111001001001101011010 is to be entered into the computer. Using octal or hexadecimal codes, this can be entered as 37111532 (octal) or 7C935A (hexadecimal). Of the two, hexadecimal codes are used much more frequently.

Programming and program execution

In most modes of usage, including use as part of intelligent instruments, computers are involved in manipulating data. This requires data values to be input, processed and output according to a sequence of operations defined by the computer program.

Programming the microprocessor within an intelligent instrument is not normally the province of the instrument user; indeed, there is rarely any provision for the user to create or modify operating programs even if he/she wished to do so. There are several reasons for this. First, the signal processing needed within an intelligent instrument is usually well defined, and therefore it is more efficient for a manufacturer to produce this rather than to have each individual user produce near-identical programs separately. Secondly, better program integrity and instrument operation are achieved if a standard program produced by the instrument manufacturer is used. Finally, use of a standard program allows it to be burnt into ROM, thereby protecting it from any failure of the instrument power supply. This also facilitates software maintenance and updates, by the mechanism of the manufacturer providing a new ROM which simply plugs into the slot previously occupied by the old ROM.

Whilst it is not normally therefore a task undertaken by the user of an intelligent instrument, some appreciation of microprocessor programming is useful background material. To illustrate the techniques involved in programming, consider a very simple program which reads in a value from a transducer, adds a prestored value to it to compensate for a bias in the transducer measurement, and outputs a corrected reading to a display device.

Let us assume that the addresses of the transducer and output display device are 00C0 and 00C1 respectively, and that the required scaling value has already been stored in memory address 0100. The instructions below are formed from the instruction set for a Z80 microprocessor and make use of CPU registers A and B.

```
IN A, C0
IN B, 100
ADD A, B
OUT C1, A
```

This list of four instructions constitutes the computer program which is necessary to execute the required task. The CPU normally executes the instructions one at a time, starting at the top of the list and working downwards (though jump and branch instructions change this order). The first instruction (IN A, C0) reads in a value from the transducer at address C0 and places the value in CPU register A (often called the accumulator). The mechanics of the execution of this instruction consist of the CPU putting the required address C0 on the address bus and then putting a command on the control bus which causes the contents of the target address (C0) to be copied onto the data bus and subsequently transferred into the A register. The next instruction (IN B, 100) reads in a value from address 100 (the prestored biasing value) and stores it in register B. The following instruction (ADD A, B) adds together the contents of registers A and B and stores the result in register A. Register A now contains the measurement read from the transducer but corrected for bias. The final instruction (OUT C1, A) transfers the contents of register A to the output device on address C1.

3.3 Interfacing

The input–output interface connects the computer to the outside world, and is therefore an essential part of the computer system. When the CPU puts the address of a peripheral onto the address bus, the input–output interface decodes the address and identifies the unique computer peripheral with which a data transfer operation is to be executed. The interface also has to interpret the command on the control bus so that the timing of the data transfer is correct. One further very important function of the input–output interface is to provide a physical electronic highway for the flow of data between the computer data bus and the external peripheral. In many computer applications, including their use within intelligent instruments, the external peripheral requires signals to be in analog form. Therefore the input–output interface must provide for conversion between these analog signals and the digital signals required by a digital computer. This is satisfied by analog-to-digital and digital-to-analog conversion elements within the input–output interface.

The design and operation of an input–output interface is a very complex subject, and a full discussion is outside the scope of this book. It is useful however to consider an overview of the input–output interface presented in simple terms.

Address decoding

A typical address bus in a microcomputer is 16 bits wide, allowing 65 536 separate

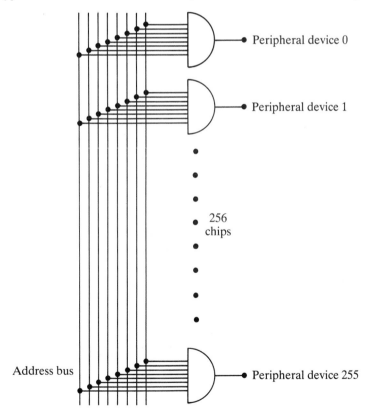

Figure 3.3   Simple hardware scheme for decoding eight address lines

addresses to be accessed in the range 0000–FFFF (in hexadecimal representation). Special commands on some computers (including the Z80) are reserved for accessing the bottom end 256 of these addresses in the range 0000–00FF, and, if these commands are used, only eight bits are needed to specify the required address. For the purpose of explaining address decoding techniques, the scheme below shows how the lower eight bits of the 16-bit address line are decoded to identify the unique address referenced by one of these special commands. Decoding of all 16 address lines follows a similar procedure but requires a substantially greater number of integrated circuit chips.

Address decoding is performed by a suitable combination of logic gates. Figure 3.3 shows a very simple hardware scheme for decoding eight address lines. This consists of 256 eight-input NAND gates, which each uniquely decode one of 256 addresses. A NAND gate is a logic element which only gives a logic level 1 output when all inputs are zero, and gives a logic level 0 output for any other combination of inputs. The inputs to the NAND gates are connected onto the lower eight lines of the address bus and the computer peripherals are connected to the output of the particular gates which decode their unique addresses. There are two pins for each input to the NAND gates which respectively invert and do not invert the input signal. By connecting the eight address lines appropriately to these two alternative pins at each input, the gate is made to decode a unique address. Consider for instance the pin connections shown in figure 3.4. This decodes address C5 (hexadecimal) which

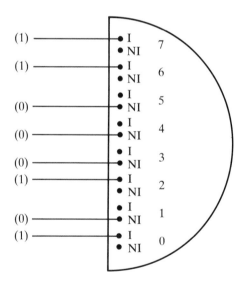

Figure 3.4   Pin connections to NAND gate to decode address C5

is 11000101 in binary. Because of the way that the input pins to the chip are connected, the NAND gate will see all zeros at its input when 11000101 is on the lower eight bits of the address bus and therefore have an output of 1. Any other binary number on the address bus will cause this NAND gate to have a zero output.

Data transfer control

The transfer of data between the computer and peripherals is managed by control and status signals carried on the control bus which determine the exact sequencing and timing of I/O operations. Such management is necessary because of the different operating speeds of the computer and its peripherals and because of the multi-tasking operation of many computers. This means that, at any particular instant when a data transfer operation is requested, either the computer or the peripheral may not be ready to take part in the transfer.

Typical control and status lines, and their meaning when set at a logic level of 1, are shown below.

BUSY	Peripheral device busy
READY	Peripheral device ready for data transfer
ENABLE	CPU ready for data transfer
ERROR	Malfunction on peripheral device

Similar control signals are set up by both the computer and peripherals, but often different conventions are used to define the status of each device. Differing conventions occur particularly when the computer and peripherals come from different manufacturers, and might mean for instance that the computer interprets a logic level of 1 as defining a device to be busy but the peripheral device uses logic level 0 to define 'device busy' on the appropriate control line. Therefore, translation of the control lines between the computer and peripherals is required, which is achieved by a further series of logic gates within the I/O interface.

Analog-to-digital (A/D) conversion

Many computer inputs, particularly those from transducers within intelligent instruments, consist of analog signals which must be converted to a digital form before they can be accepted by the computer. This conversion is performed by an analog-to-digital conversion circuit within the computer interface.

Important factors in the design of an analog-to-digital converter are the speed of conversion and the number of digital bits used to represent the analog signal level. The minimum number of bits used in analog-to-digital converters is eight, which means that the analog signal can be represented to a resolution of one part in 256 if the input signal is carefully scaled to make full use of the converter range. It is more common however to use either 10-bit or 12-bit analog-to-digital converters which give resolutions respectively of one part in 1024 and one part in 4096. Several types of analog-to-digital converter exist. These differ in the technique used to effect signal conversion, in operational speed, and in cost.

The simplest type of analog-to-digital converter is the *counter analog-to-digital converter*, as shown in figure 3.5. This, like most types of analog-to-digital converter, does not convert continuously, but in a stop–start mode triggered by special signals on the computer's control bus. At the start of each conversion cycle, the counter is set to zero. The digital counter value is converted to an analog signal by a digital-to-analog converter (a discussion of digital-to-analog converters follows in the next section), and a comparator then compares this analog counter value with the

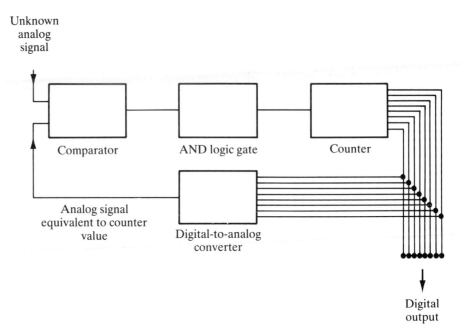

Figure 3.5   Counter analog-to-digital converter

unknown analog signal. The output of the comparator forms one of the inputs to an AND logic gate. The other input to the AND gate is a sequence of clock pulses. The comparator acts as a switch which can turn on and off the passage of pulses from the clock through the AND gate. The output of the AND gate is connected to the input of the digital counter. Following reset of the counter at the start of the conversion cycle, clock pulses are applied continuously to the counter through the AND gate, and the analog signal at the output of the digital-to-analog converter gradually increases in magnitude. At some point in time, this analog signal becomes equal in magnitude to the unknown signal at the input to the comparator. The output of the comparator changes state in consequence, closing the AND gate and stopping further increments of the counter. At this point in time, the value held in the counter is a digital representation of the level of the unknown analog signal.

Digital-to-analog (D/A) conversion

Digital-to-analog conversion is much simpler to achieve than analog-to-digital conversion and the cost of building the necessary hardware circuit is considerably less. It is required wherever the output of an intelligent instrument needs to be presented on a display device which operates in an analog manner. A common form of digital-to-analog converter is illustrated in figure 3.6. This is shown with eight bits for simplicity of explanation, although in practice 10- and 12-bit D/A converters are used more frequently.

This form of D/A converter consists of a resistor ladder network on the input to an operational amplifier. The analog output voltage from the amplifier is given by:

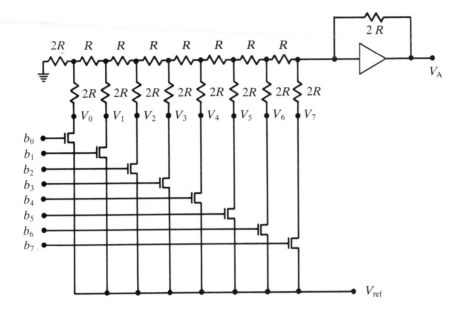

Figure 3.6   Common form of digital-to-analog converter

$$V_A = V_7 + \frac{V_6}{2} + \frac{V_5}{4} + \frac{V_4}{8} + \frac{V_3}{16} + \frac{V_2}{32} + \frac{V_1}{64} + \frac{V_0}{128}$$

$V_0, \ldots, V_7$ are set at either the reference voltage level V_{ref} or at zero volts according to whether an associated switch is open or closed. Each switch is controlled by the logic level of one of the bits 0–7 of the eight-bit binary signal being converted. A particular switch is open if the relevant binary bit has a value of 0 and closed if the value is 1.

Consider for example a digital signal with binary value of 11010100. The values of $V_7, \ldots, V_0$ are therefore:

$$V_7 = V_6 = V_4 = V_2 = V_{ref} \qquad V_5 = V_3 = V_1 = V_0 = 0$$

The analog output from the converter is then given by:

$$V_A = V_{ref} + \frac{V_{ref}}{2} + \frac{V_{ref}}{8} + \frac{V_{ref}}{32}$$

Other interface considerations

The foregoing discussion has presented some of the necessary elements in an input–output interface in a relatively simplistic manner which is just sufficient to give the reader the flavor of what is involved in an interface. Much fine detail has been omitted, and the amount of work involved in the practical design of a real interface should not be underestimated.

One significant omission so far is discussion of the scaling which is generally required within the analog–digital interface of a computer. The raw analog input and output signals are generally either too large or too small for compatibility with the operating voltage levels of a digital computer and they have to be scaled upwards or downwards. This is normally achieved by operational amplifiers and/or potentiometers.

The main features of an operational amplifier are its high gain (typically ×1000 000) and large bandwidth (typically 1 MHz or better). When used at high frequencies, however, the bandwidth becomes significant. The quality of an amplifier is often measured by a criterion called the gain–bandwidth product, which is the product of its gain and bandwidth. Other important attributes of the operational amplifier, particularly when used within intelligent instruments, are its distortion level, overload recovery capacity and offset level. Special versions of operational amplifiers which are particularly good in these attributes have been developed for instrumentation applications.

3.4   Communications

Serious problems can arise when several intelligent instruments are required to communicate with a supervisory computer in a plant computer control scheme. In such cases, the intelligent instruments often come from several different manufacturers who use a variety of parallel interface protocols. Some examples of interface protocols are the IEE488 bus, S-100 bus, CAMAC, MEDIA and HDLC. Each of these has a different interrupt structure, data timing system, physical connection mechanism, operating voltage level and programming requirement. Therefore, connecting an intelligent instrument and host computer using different interface protocols requires considerable expertise and effort.

One way of interconnecting a host computer and several intelligent instruments is to connect each instrument directly to the host computer by its own signal cable in a configuration known as a *star network*. An alternate form of network is the *ring network*, and this makes the process of connecting different computer systems easier by providing a common ring communication protocol. Two well known ring protocols are *Ethernet* and the *Cambridge Ring*, and most computer and intelligent instrument manufacturers provide standard conversion modules which allow their equipment to interface to one of these standard ring networks. The ring network also has an additional advantage in using a ring of cable (often simple coaxial cable), which avoids the expense of separate cables from the host computer to each instrument. In certain circumstances, ring networks can impose an unacceptable delay on the transport of data to and from the host computer and can thus degrade the quality of process control. This problem is not common, however.

Mention must also be made of current efforts to rationalize the diversity of communication protocols available. This started with the publication in 1978 by the International Standards Organization of the *Open Systems Interconnection* seven-layer model (ISO-7), covering all aspects of computer communications. The *Manufacturing Automation Protocol* (MAP), conceived by General Motors in 1980, conforms with ISO-7, and is an attempt at providing a standard computer communications protocol. MAP is now being adopted by many large companies and organizations, including the European ESPRIT research program, and promises to realize what it set out to achieve, that is, the provision of an industry-wide common communications standard.

It should also be realized that communication problems can be avoided almost entirely by using a serial communication protocol such as RS232 instead of parallel communication. Serial communication transmits data down a single line with special markers to define the start and finish of each data stream, as opposed to parallel communication where data are transmitted n bits at a time along n parallel data lines. Serial communication is, however, much slower than even a ring network and is therefore limited in its applicability.

Making the reader aware of these potential communication difficulties is as far as this text can go, as the communications aspects of interfacing different computer systems is a highly specialized subject which is well out of the scope of this book. The reader so interested is directed to the book of Barney (1985).

3.5   Intelligent instruments in use

The intelligent instrument behaves as a black box as far as the user is concerned, and no knowledge of its internal mode of operation is required in normal measurement situations. The foregoing discussion on the operating principles of intelligent instruments is therefore presented for interest only and is not mandatory reading for potential users of such instruments.

Intelligent instruments offer many advantages over their non-intelligent counterparts, principally because of the improvement in accuracy achieved by processing the output of transducers to correct for errors inherent in the measurement process. Proper procedures must always be followed in their use, however, as discussed in section 6.5, to avoid the possibility of introducing extra sources of measurement error. The solutions which intelligent instruments offer to many problems occurring in measurement systems will be discussed at various points in the chapters following in this first part of the book on measurement principles, and particular applications will be presented later in part B.

One such example of the benefit that intelligence can bring to instruments is in volume flow-rate measurement, where the flow rate is inferred by measuring the differential pressure across an orifice plate placed in a fluid-carrying pipe (see chapter 16 on flow measurement for more details). The flow rate is proportional to the square root of the difference in pressure across the orifice plate. For a given flow rate, this relationship is affected both by the temperature and by the mean pressure in the pipe, and changes in the ambient value of either of these cause measurement errors.

A typical intelligent flow-rate-measuring instrument contains three transducers, a primary one measuring pressure difference across an orifice plate and secondary ones measuring absolute pressure and temperature. The instrument is programmed to correct the output of the primary differential pressure transducer according to the values measured by the secondary transducers, using appropriate physical laws which quantify the effect of ambient temperature and pressure changes on the fundamental relationship between flow and differential pressure. The instrument is also normally programmed to convert the square-root relationship between flow and signal output into a direct one, making the output much easier to interpret. Typical accuracy levels of such intelligent flow-measuring instruments are ±0.1%, compared with ±0.5% for their non-intelligent equivalents, showing a factor of 5 improvement.

Intelligent instruments usually provide many other facilities in addition to those mentioned above such as:

(i)   signal damping with selectable time constants;
(ii)   switchable ranges (using several primary transducers within the instrument which each measure over a different range);
(iii)   switchable output units (e.g. display in Imperial or SI units);
(iv)   diagnostic facilities;
(v)   remote adjustment and control of instrument options from up to 1500 m away via four-way, 20 mA signal lines.

Exercises

3.1   Describe briefly the three essential elements of a microcomputer system.

3.2   Write down the binary, octal and hexadecimal representations of the following decimal numbers: (a) 57, (b) 101, (c) 175, (d) 259, (e) 999, (f) 1234.

3.3   Write down the one's complement and two's complement eight-bit representations of the following decimal numbers: (a) 47, (b) −119, (c) −101, (d) 86, (e) 108.

4

Bridge Circuits

Bridge circuits are used very commonly as a variable conversion element in measurement systems and produce an output in the form of a voltage level which changes as the measured physical quantity changes in value. They provide an accurate method of measuring resistance, inductance and capacitance values, and enable very small changes in these quantities about a nominal value to be detected. They are of immense importance in measurement system technology because so many transducers measuring physical quantities have an output which is expressed as a change in resistance, inductance or capacitance. The displacement-measuring strain gauge, which has a varying-resistance output, is but one example of this class of transducers. Excitation of the bridge is normally by a d.c. voltage for resistance measurement and by an a.c. voltage for inductance or capacitance measurement. Both null and deflection types of bridge exist, and, in a like manner to instruments in general, null types are mainly employed for calibration purposes and deflection types used within closed-loop automatic control schemes.

4.1   Wheatstone bridge (null-type, d.c. excitation)

The standard Wheatstone bridge is by far the most commonly known bridge and has the form shown in figure 4.1. The four arms of the bridge consist of the unknown resistance R_u, two equal-value resistors R_2 and R_3 and a variable resistor R_v in the form of a decade resistance box. A d.c. voltage V_i is applied across the points AC and the resistance R_v is varied until the voltage measured across points BD is zero. This null point is usually measured with a high-sensitivity galvanometer.

To analyze the Wheatstone bridge, define the current flowing in each arm to be $I_1, \ldots, I_4$ as shown in figure 4.1. Normally, if a high-impedance voltage-measuring instrument is used, the current I_m drawn by the measuring instrument will be very small and can be approximated to zero. If this assumption is made, then:

$$\text{for } I_m = 0, \qquad I_1 = I_3 \qquad \text{and} \qquad I_2 = I_4$$

Looking at path ADC, we have a voltage V_i applied across a resistance $R_u + R_3$ and by Ohm's law:

$$I_1 = \frac{V_i}{R_u + R_3}$$

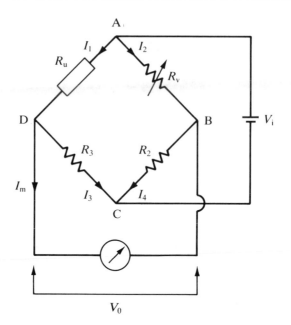

Figure 4.1   Analysis of Wheatstone bridge

Similarly for path ABC,

$$I_2 = \frac{V_i}{R_v + R_2}$$

Now we can calculate the voltage drop across AD and AB:

$$V_{AD} = I_1 R_u = \frac{V_i R_u}{R_u + R_3}$$

$$V_{AB} = I_2 R_v = \frac{V_i R_v}{R_v + R_2}$$

By the principle of superposition,

$$V_0 = V_{BD} = V_{BA} + V_{AD} = -V_{AB} + V_{AD}$$

Thus

$$V_0 = -\frac{V_i R_v}{R_v + R_2} + \frac{V_i R_u}{R_u + R_3} \qquad (4.1)$$

At the null point $V_0 = 0$, so:

$$\frac{R_u}{R_u + R_3} = \frac{R_v}{R_v + R_2}$$

Inverting both sides:

$$\frac{R_u + R_3}{R_u} = \frac{R_v + R_2}{R_v}$$

i.e.

$$\frac{R_3}{R_u} = \frac{R_2}{R_v}$$

or

$$R_u = \frac{R_3 R_v}{R_2} \tag{4.2}$$

Thus, if $R_2 = R_3$, then $R_u = R_v$.

As R_v is an accurately known value because it is derived from a variable decade resistance box, this means that R_u is also accurately known.

4.2 Deflection-type d.c. bridge

A deflection-type bridge with d.c. excitation is shown in figure 4.2. This differs from the Wheatstone bridge mainly in that the variable resistance R_v is replaced by a fixed resistance R_1 of the same value as the nominal value of the unknown resistance R_u. As the resistance R_u changes, so the output voltage V_0 varies, and this relationship between V_0 and R_u must be calculated.

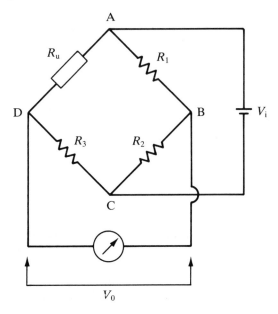

Figure 4.2   Deflection-type d.c. bridge

We start again by assuming that a high-impedance voltage-measuring instrument is used and the current drawn by it, I_m, can be approximated to zero (the case when this assumption does not hold is covered later in chapter 6).

The analysis is exactly the same as for the preceding example of the Wheatstone bridge, except that R_v is replaced by R_1. Thus, from equation (4.1), we have:

$$V_0 = V_i \left(\frac{R_u}{R_u + R_3} - \frac{R_1}{R_1 + R_2} \right) \quad (4.3)$$

When R_u is at its nominal value, i.e. for $R_u = R_1$, it is clear that $V_0 = 0$ (since $R_2 = R_3$). For other values of R_u, V_0 has negative and positive values which vary in a non-linear way with R_u.

Such a non-linear relationship between output reading and measured quantity is unfortunate and does not conform with our normal requirement for a linear input–output relationship. The method of coping with this non-linearity varies according to the form of primary transducer involved in the measurement system.

One special case is where the change in the unknown resistance R_u is typically small compared with the nominal value of R_u. If we calculate the new voltage V_0' when the resistance R_u in equation (4.3) changes by an amount ΔR_u, we have:

$$V_0' = V_i \left(\frac{R_u + \Delta R_u}{R_u + \Delta R_u + R_3} - \frac{R_1}{R_i + R_2} \right) \quad (4.4)$$

The change of voltage output is therefore given by:

$$\Delta V_0 = V_0' - V_0 = \frac{V_i \, \Delta R_u}{R_u + \Delta R_u + R_3}$$

If $\Delta R_u \ll R_u$, then the following linear relationship is obtained:

$$\frac{\Delta V_0}{\Delta R_u} = \frac{V_i}{R_u + R_3} \quad (4.5)$$

This expression describes the measurement sensitivity of the bridge. Such an approximation to make the relationship linear is valid for transducers such as strain gauges where the typical changes of resistance with strain are very small compared with the nominal gauge resistance.

However, many instruments which are inherently linear themselves at least over a limited measurement range, such as resistance thermometers, exhibit large changes in output as the input quantity changes, and the approximation of equation (4.5) cannot be applied. In such cases, specific action must be taken to improve linearity in the relationship between the bridge output voltage and the measured quantity. One common solution to this problem is to make the values of the resistances R_2 and R_3 at least 10 times those of R_1 and R_u (nominal). The effect of this is best observed by looking at a numerical example.

Consider a platinum resistance thermometer with a range of 0–50°C, whose resistance at 0°C is 500 Ω and whose resistance varies with temperature at the rate of

4 $\Omega/°C$. Over this range of measurement, the output characteristic of the thermo-meter itself is nearly perfectly linear. (*N.B.* The subject of resistance thermometers is discussed further in chapter 15.)

Taking first the case where $R_1 = R_2 = R_3 = 500\ \Omega$ and $V_i = 10\ V$, and applying equation (4.3):

at 0°C

$$V_0 = 0$$

at 25°C

$$R_u = 600\ \Omega \qquad V_0 = 10 \left(\frac{600}{1100} - \frac{500}{1000} \right) = 0.455\ V$$

at 50°C

$$R_u = 700\ \Omega \qquad V_0 = 10 \left(\frac{700}{1200} - \frac{500}{1000} \right) = 0.833\ V$$

This relationship between V_0 and R_u is plotted as curve A in figure 4.3 and the non-linearity is apparent. Inspection of the manner in which the output voltage V_0 above changes for equal steps of temperature change also clearly demonstrates the non-linearity:

for the temperature change from 0 to 25°C, the change in V_0 is $(0.455 - 0) = 0.455\ V$

for the temperature change from 25 to 50°C, the change in V_0 is
$$(0.833 - 0.455) = 0.378\ V$$

If the relationship was linear, the change in V_0 for the 25 to 50°C temperature step would also be 0.455 V, giving a value for V_0 of 0.910 V at 50°C.

Now take the case where $R_1 = 500\ \Omega$ but $R_2 = R_3 = 5000\ \Omega$ and let $V_i = 26.1\ V$:

at 0°C

$$V_0 = 0$$

at 25°C

$$R_u = 600\ \Omega \qquad V_0 = 26.1 \left(\frac{600}{5600} - \frac{500}{5500} \right) = 0.424\ V$$

at 50°C

$$R_u = 700\ \Omega \qquad V_0 = 26.1 \left(\frac{700}{5700} - \frac{500}{5500} \right) = 0.833\ V$$

This relationship is shown as curve B in figure 4.3 and a considerable improvement

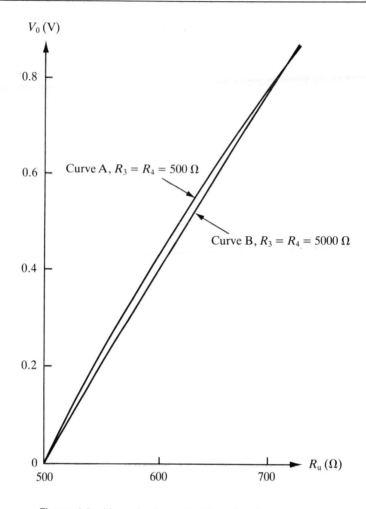

Figure 4.3   Linearization of bridge circuit characteristic

in linearity is achieved. This is more apparent if the differences in values for V_0 over the two temperature steps are inspected:

from 0 to 25°C, the change in V_0 is 0.424 V

from 25 to 50°C, the change in V_0 is 0.409 V

The changes in V_0 over the two temperature steps are much closer to being equal than before, demonstrating the improvement in linearity.

In increasing the values of R_2 and R_3, it was also necessary to increase the excitation voltage from 10 V to 26.1 V to obtain the same output levels. In practical applications, V_i would normally be set at the maximum level consistent with the limitation of the effect of circuit heating in order to maximize the measurement sensitivity ($\Delta V_0/\Delta R_u$ relationship). It would therefore not be possible to increase V_i further if R_2 and R_3 were increased, and the general effect of such an increase in R_2 and R_3 is thus a decrease in the sensitivity of the measurement system.

The importance of this inherent non-linearity in the bridge output relationship is greatly diminished if the primary transducer and bridge circuit are incorporated as elements within an intelligent instrument. In that case, digital computation is applied to produce an output in terms of the measured quantity which automatically compensates for the non-linearity in the bridge circuit.

Example

A certain type of pressure transducer, designed to measure pressures in the range 0–10 bar, consists of a diaphragm with a strain gauge cemented to it to detect diaphragm deflections. The strain gauge has a nominal resistance of 120 Ω and forms one arm of a Wheatstone bridge circuit, with the other three arms each having a resistance of 120 Ω. The bridge output is measured by an instrument whose input impedance can be assumed infinite. If, in order to limit heating effects, the maximum permissible gauge current is 30 mA, calculate the maximum permissible bridge excitation voltage. If the sensitivity of the strain gauge is 338 mΩ/bar and the maximum bridge excitation voltage is used, calculate the bridge output voltage when measuring a pressure of 10 bar.

Solution

This is the type of bridge circuit shown in figure 4.2 in which the components have the following values:

$$R_1 = R_2 = R_3 = 120 \, \Omega$$

Defining I_1 to be the current flowing in path ADC of the bridge, we can write:

$$V_i = I_1(R_u + R_3)$$

At balance, $R_u = 120 \, \Omega$ and the maximum value allowable for I_1 is 0.03 A. Hence:

$$V_i = 0.03(120 + 120) = 7.2 \text{ V}$$

Thus the maximum bridge excitation voltage allowable is 7.2 V.
For a pressure of 10 bar applied, the resistance change is 3.38 Ω, i.e. R_u is then equal to 123.38 Ω. Applying equation (4.3), we can write:

$$V_0 = V_i \left(\frac{R_u}{R_u + R_3} - \frac{R_1}{R_1 + R_2} \right) = 7.2 \left(\frac{123.38}{243.38} - \frac{120}{240} \right) = 50 \text{ mV}$$

Thus, if the maximum permissible bridge excitation voltage is used, the output voltage is 50 mV when a pressure of 10 bar is measured.

4.3   Error analysis

In the application of bridge circuits, and most especially when they are being used for calibration purposes, the contribution of component-value tolerances to total measurement-system accuracy limits must be clearly understood. The maximum measurement error is determined by first finding the value of R_u in equation (4.2) with each parameter in the equation set at that limit of its tolerance which produces the maximum value of R_u. Similarly, the minimum possible value of R_u is calculated, and the required error band is then the span between these maximum and minimum values.

Example

In the Wheatstone bridge circuit of figure 4.1, R_v is a decade resistance box of accuracy $\pm 0.2\%$ and $R_2 = R_3 = 500\ \Omega \pm 0.1\%$. If the value of R_v at the null position is 520.4 Ω, determine the error band for R_u expressed as a percentage of its nominal value.

Solution

Applying equation (4.2) with $R_v = 520.4\ \Omega + 0.2\% = 521.44\ \Omega$, $R_3 = 5000\ \Omega + 0.1\% = 5005\ \Omega$, $R_2 = 5000\ \Omega - 0.1\% = 4995\ \Omega$, we get:

$$R_v = \frac{521.44 \times 5005}{4995} = 522.48\ \Omega \qquad (= +0.4\%)$$

Applying equation (4.2) with $R_v = 520.4\ \Omega - 0.2\% = 519.36\ \Omega$, $R_3 = 5000\ \Omega - 0.1\% = 4995\ \Omega$, $R_2 = 5000\ \Omega + 0.1\% = 5005\ \Omega$, we get:

$$R_v = \frac{519.36 \times 4995}{5005} = 518.32\ \Omega \qquad (= -0.4\%)$$

Thus the error band for R_u is $\pm 0.4\%$.

The cumulative effect of errors in individual bridge circuit components is clearly seen. Although the maximum error in any one component is $\pm 0.2\%$, the possible error in the measured value of R_u is $\pm 0.4\%$. Such a magnitude of error is often not acceptable, and special measures are taken to overcome the introduction of error by component-value tolerances. One such practical measure is the introduction of apex balancing. This is one of many methods of bridge balancing which all produce a similar result.

Apex balancing

One form of apex balancing consists of placing an additional variable resistor R_5 at the junction C between the resistances R_2 and R_3, and applying the excitation voltage V_i to the wiper of this variable resistance, as shown in figure 4.4.

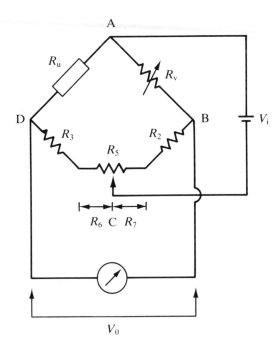

Figure 4.4   Apex balancing

For calibration purposes, R_u and R_v are replaced by two equal resistances whose values are accurately known, and R_5 is varied until the output voltage V_0 is zero. At this point, if the portions of resistance on either side of the wiper on R_5 are R_6 and R_7 (such that $R_5 = R_6 + R_7$), we can write:

$$R_3 + R_6 = R_2 + R_7$$

We have thus eliminated any source of error due to the tolerance of R_2 and R_3, and the error in the measured value of R_u depends only on the accuracy of one component, the decade resistance box R_v.

4.4   a.c. bridges

Bridges with a.c. excitation are used to measure unknown impedances. As for d.c. bridges, both null and deflection types exist, with null types being generally reserved for calibration duties.

Null-type impedance bridge

A typical null-type impedance bridge is shown in figure 4.5. The null point can be conveniently detected by monitoring the output with a pair of headphones connected

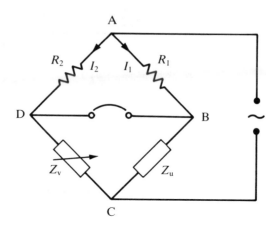

Figure 4.5   Null-type impedance bridge

via an operational amplifier across the points BD. This is a much cheaper method of null detection than the application of an expensive galvanometer which is required for a d.c. Wheatstone bridge. Referring to figure 4.5:

at the null point

$$I_1 R_1 = I_2 R_2$$
$$I_1 Z_u = I_2 Z_v$$

Thus:

$$Z_u = \frac{Z_v R_1}{R_2} \tag{4.6}$$

If Z_u is capacitive, i.e.

$$Z_u = R_u + \frac{1}{j\omega C_u}$$

then Z_v must consist of a variable-resistance box and a variable-capacitance box in series, both of which are readily available.

If Z_u is inductive, then

$$Z_u = R_u + j\omega L_u$$

In this case, Z_v must consist of a variable-resistance box and a variable-inductance box. However, the latter are not so readily available because it is difficult and hence expensive to manufacture a set of fixed-value inductors to make up a variable-inductance box. For this reason, an alternative type of bridge circuit known as the Maxwell bridge is used to measure unknown inductances.

Maxwell bridge

A Maxwell bridge is shown in figure 4.6. The requirement for a variable-inductance box is avoided by introducing instead a second variable resistance. The circuit

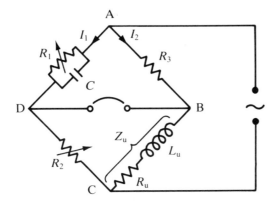

Figure 4.6   Maxwell bridge

requires one standard fixed-value capacitor, two variable-resistance boxes and one standard fixed-value resistor, all of which are readily available and cheap components. Referring to figure 4.6, we have at the null-output point:

$$I_1 Z_{AD} = I_2 Z_{AB}$$
$$I_1 Z_{DC} = I_2 Z_{AD}$$

Thus

$$\frac{Z_{BC}}{Z_{AB}} = \frac{Z_{DC}}{Z_{AD}}$$

or

$$Z_{BC} = \frac{Z_{DC} Z_{AB}}{Z_{AD}} \tag{4.7}$$

The quantities in equation (4.7) have the following values:

$$\frac{1}{Z_{AD}} = \frac{1}{R_1} + j\omega C \quad \text{or} \quad Z_{AD} = \frac{R_1}{1 + j\omega C R_1}$$

$$Z_{AB} = R_3 \qquad Z_{BC} = R_u + j\omega L_u \qquad Z_{DC} = R_2$$

Substituting the values into equation (4.7):

$$R_u + j\omega L_u = \frac{R_2 R_3 (1 + j\omega C R_1)}{R_1}$$

Taking real and imaginary parts:

$$R_u = \frac{R_2 R_3}{R_1} \qquad L_u = R_2 R_3 C \qquad (4.8)$$

Example

In the Maxwell bridge shown in figure 4.6, let the fixed-value bridge components have the following values:

$$R_3 = 200\,\Omega \qquad C = 0.1\,\mu F$$

Calculate the value of the unknown impedance (L_u, R_u) if $R_1 = 500\,\Omega$ and $R_2 = 100\,\Omega$ at balance.

Solution

Substituting values into the relations developed in equation (4.8) above:

$$R_u = \frac{R_2 R_3}{R_1} = \frac{100 \times 200}{500} = 40\,\Omega$$

$$L_u = R_2 R_3 C = \frac{100 \times 200 \times 1}{10^7} = \frac{2}{10^3} = 20\,mH$$

Deflection-type a.c. bridge

A common deflection type of a.c. bridge circuit is shown in figure 4.7. For capacitance measurement:

$$Z_u = \frac{1}{j\omega C_u} \qquad Z_1 = \frac{1}{j\omega C_1}$$

For inductance measurement:

$$Z_u = j\omega L_u \qquad Z_1 = j\omega L_1$$

Analysis of the circuit to find the relationship between V_0 and Z_u is greatly simplified if one assumes that I_m is very small. For $I_m \simeq 0$, currents in the two branches of the bridge, as defined in figure 4.7, are given by:

$$I_1 = \frac{V_s}{Z_1 + Z_u} \qquad I_2 = \frac{V_s}{R_2 + R_3}$$

Also

$$V_{AC} = I_1 Z_u \qquad \text{and} \qquad V_{BC} = I_2 R_3$$

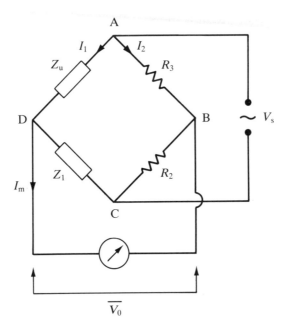

Figure 4.7   Deflection-type a.c. bridge

Then

$$V_0 = V_{AB} = V_{AC} - V_{BC} = V_s \left(\frac{Z_u}{Z_1 + Z_u} - \frac{R_3}{R_2 + R_3} \right)$$

Thus for capacitances:

$$V_0 = V_s \left(\frac{1/C_u}{1/C_1 + 1/C_u} - \frac{R_3}{R_2 + R_3} \right) = V_s \left(\frac{C_1}{C_1 + C_u} - \frac{R_3}{R_2 + R_3} \right) \qquad (4.9)$$

and for inductances:

$$V_0 = V_s \left(\frac{L_u}{L_1 + L_u} - \frac{R_3}{R_2 + R_3} \right) \qquad (4.10)$$

This latter relationship (4.10) is in practice only approximate since inductive impedances are never pure inductances as assumed but always contain a finite resistance (i.e. $Z_u = j\omega L_u + R$).

Exercises

4.1   In the Wheatstone bridge shown in figure 4.8, the resistive components have the following nominal values: $R_1 = 3$ kΩ, $R_2 = 6$ kΩ, $R_3 = 8$ kΩ, $R_4 = 4$ kΩ. The actual value of each resistance is related to the nominal value according to $R_{\text{actual}} = R_{\text{nominal}} + \Delta R$ where ΔR has the following values: $\Delta R_1 = 30\ \Omega$, $\Delta R_2 = -20\ \Omega$, $\Delta R_3 = 40\ \Omega$, $\Delta R_4 = -50\ \Omega$. Calculate the open-circuit bridge output voltage if the bridge supply voltage V_s is 50 V.

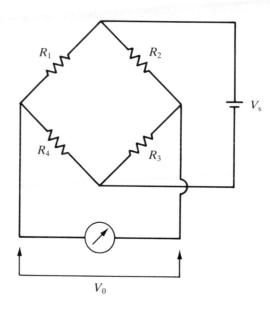

Figure 4.8   Wheatstone bridge (exercise 4.1)

4.2   (a) Figure 4.2 shows a d.c. bridge circuit designed to measure the value of an unknown resistance R_u. The impedance of the instrument measuring the output voltage V_0 is such that it draws negligible current. Derive an expression for the output voltage V_0 in terms of the input voltage V_i and the four resistances in the bridge, R_u, R_1, R_2 and R_3.
(b) Suppose that the unknown resistance R_u is a resistance thermometer whose resistance at 100°C is 500 Ω and whose resistance varies with temperature at the rate of 0.5 Ω/°C for small temperature changes around 100°C. Calculate the sensitivity of the total measurement system for small changes in temperature around 100°C, given the following resistance and voltage values measured at 15°C by instruments calibrated at 15°C: $R_1 = 500\ \Omega$, $R_2 = R_3 = 5000\ \Omega$, $V_i = 10$ V.
(c) If the resistance thermometer is measuring a fluid whose true temperature is 104°C, calculate the error in the indicated temperature if the ambient temperature around the bridge circuit is 20°C instead of the calibration temperature of 15°C, given the following additional information: Voltage-measuring instrument zero drift coefficient = +1.3 mV/°C. Voltage-measuring instrument sensitivity drift coefficient = 0. Resistances R_1, R_2 and R_3 have a positive temperature coefficient of +0.2% of nominal value per °C. Voltage source V_i is unaffected by temperature changes.

4.3   Four strain gauges of resistance 120 Ω each are arranged into a Wheatstone bridge configuration such that each of the four arms in the bridge has one strain gauge in it. The maximum permissible current in each strain gauge is 100 mA. What is the maximum bridge supply voltage allowable, and what power is dissipated in each strain gauge with that supply voltage?

4.4   In the d.c. bridge circuit shown in figure 4.9, the resistive components have the following values: $R_1 = R_2 = R_3 = R_4 = 120\ \Omega$, $R_A = R_P = 1000\ \Omega$. What is the resistance of each of the left-hand arms of the bridge when the wiper (sliding element) is at the center

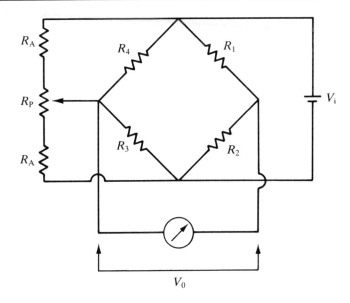

Figure 4.9   D.c. bridge (exercise 4.4)

of the potentiometer (R_P), and what percentage variation of the ratio of these two arms can be achieved by moving the potentiometer wiper?

4.5   A Maxwell bridge, designed to measure the unknown impedance (R_u, L_u) of a coil, is shown in figure 4.6.
(a) Derive an expression for R_u and L_u under balance conditions.
(b) Show how the scale of R_1 could be calibrated to indicate the Q value of the coil.

4.6   A Hays bridge is often used for measuring the inductance of high-Q coils and has the configuration shown in figure 4.10.
(a) Obtain the bridge balance conditions.
(b) Show that if the Q value of an unknown inductor is high, the expression for the inductance value when the bridge is balanced is independent of frequency.
(c) If the Q value is high, calculate the value of the inductor if the bridge component values at balance are as follows: $R_2 = R_3 = 1000\ \Omega$, $C = 0.02\ \mu F$.

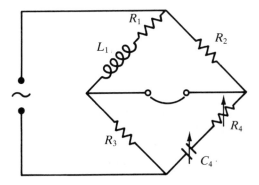

Figure 4.10   Hays bridge (exercise 4.6)

5

Signal Processing

Signal processing is concerned with improving the quality of the reading or signal at the output of a measurement system. The form which signal processing takes depends on the nature of the raw output signal from a measurement transducer. Procedures of signal amplification, signal attenuation, signal linearization, bias removal and signal filtering are applied according to the form of correction required in the raw signal.

The implementation of signal processing procedures can be carried out either by analog techniques or by digital computation. For the purposes of explaining the procedures involved, this chapter mainly describes analog signal processing and concludes with a relatively brief discussion of the equivalent digital signal processing techniques. One particular reason for this method of treatment is that some prior analog signal conditioning is often necessary even when the major part of the signal processing is carried out digitally.

Choice between analog and digital signal processing is largely determined by the degree of accuracy required in the signal processing procedure. Digital processing is very much more accurate than analog processing but the cost of the equipment involved is much greater and the processing time is much longer. Normal practice, therefore, is to use analog techniques for all signal processing tasks except where the accuracy of this is insufficient. It should be noted also that in some measurement situations where a physical quantity is measured by an inherently inaccurate transducer, the extra accuracy provided by digital signal processing is insignificant and therefore such expensive and slow processing techniques are inappropriate.

5.1 Signal amplification

Signal amplification is carried out when the typical signal output level of a measurement transducer is considered to be too low. Amplification by analog means is carried out by an operational amplifier. This is normally required to have a high input impedance so that its loading effect on the transducer output signal is minimized. In some circumstances, such as when amplifying the output signal from accelerometers and some optical detectors, the amplifier must also have a high-frequency response, to avoid distortion of the measured quantity–output reading relationship.

The operational amplifier is an electronic device which has two input terminals

60

Figure 5.1   Operational amplifier connected for signal amplification

and one output terminal, the two inputs being known as the inverting input and non-inverting input respectively. When used for signal processing duties, it is connected in the configuration shown in figure 5.1. The raw (unprocessed) signal V_i is connected to the inverting input through a resistor R_1 and the non-inverting input is connected to ground. A feedback path is provided from the output terminal through a resistor R_2 to the inverting input terminal. The processed signal V_0 at the output terminal is then related to the voltage V_i at the input terminal by the expression (assuming ideal operational amplifier characteristics):

$$V_0 = -\frac{R_2}{R_1}V_i \qquad (5.1)$$

The amount of signal amplification is therefore defined by the relative values of R_1 and R_2. This ratio between R_1 and R_2 in the amplifier configuration is often known as the amplifier gain. If, for instance, $R_1 = 1\ M\Omega$ and $R_2 = 10\ M\Omega$, an amplification factor of 10 is obtained (i.e. gain = 10). It is important to note that in this standard way of connecting the operational amplifier, the sign of the processed signal is inverted. This can be corrected for if necessary by feeding the signal through a further amplifier set up for unity gain ($R_1 = R_2$). This inverts the signal again and returns it to its original sign.

5.2   Signal attenuation

One method of attenuating signals by analog means is to use a potentiometer connected in a voltage-dividing circuit, as shown in figure 5.2. For the potentiometer wiper positioned a distance of X_w along the resistance element of total length X_t, the voltage level of the processed signal V_0 is related to the voltage level of the raw signal V_i by the expression:

$$V_0 = \frac{X_w}{X_t}V_i$$

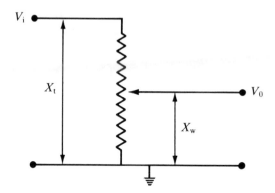

Figure 5.2   Potentiometer in voltage-dividing circuit

An alternative device to the potentiometer for signal attenuation is the operational amplifier. This is connected in exactly the same way as for an amplifier as shown in figure 5.1, but R_1 is chosen to be greater than R_2. Equation (5.1) still holds and therefore if R_1 is chosen to be 10 MΩ and R_2 as 1 MΩ, an attenuation factor of 10 is achieved (gain = 0.1). Use of an operational amplifier as an attenuating device is a more expensive solution than using a potentiometer, but it has advantages in terms of its smaller size and low power consumption.

5.3   Signal linearization

Several types of transducer used in measuring instruments have an output which is a non-linear function of the measured quantity input. In many cases, this non-linear signal can be converted to a linear one by special operational amplifier configurations which have an equal and opposite non-linear relationship between the amplifier input and output terminals.

For example, light intensity transducers typically have an exponential relationship between the output signal and the input light intensity, i.e.:

$$V_0 = Ke^{-\alpha Q} \tag{5.2}$$

where Q is the light intensity, V_0 is the voltage level of the output signal, and K and α are constants.

If a diode is placed in the feedback path between the input and output terminals of the amplifier as shown in figure 5.3, the relationship between the amplifier output voltage V_2 and input voltage V_1 is given by:

$$V_2 = C \log_e(V_1) \tag{5.3}$$

If the output of the light transducer with characteristic given by equation (5.2)

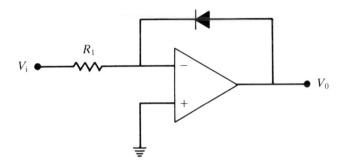

Figure 5.3   Operational amplifier connected for signal linearization

is conditioned by an amplifier of characteristic given by equation (5.3), the voltage level of the processed signal is given by:

$$V_2 = C \log(K) - \alpha C Q \qquad (5.4)$$

Expression (5.4) shows that the output signal now varies linearly with light intensity Q but with an offset of $C \log(K)$. This offset would normally be removed by further signal conditioning, as described below.

5.4   Bias removal

Sometimes, either because of the nature of the measurement transducer itself, or as a result of other signal conditioning operations (see section 5.3), a bias exists in the output signal. This can be expressed mathematically for a physical quantity x and measurement signal y as:

$$y = Kx + C \qquad (5.5)$$

where C represents a bias in the output signal which needs to be removed by signal processing. Analog processing consists of using an operational amplifier connected in a differential amplification mode, as shown in figure 5.4. Referring to this circuit, for $R_1 = R_2$ and $R_3 = R_4$, the output V_0 is given by:

$$V_0 = (R_3/R_1)(V_c - V_i) \qquad (5.6)$$

where V_i is the unprocessed signal y equal to $(Kx + C)$ and V_c is the output voltage from a potentiometer supplied by a known reference voltage V_{ref}, which is set such that $V_c = C$. Now, substituting these values for V_i and V_c into equation (5.6) and referring the quantities back into equation (5.5), we find that:

$$y = K'x \qquad (5.7)$$

Figure 5.4   Operational amplifier connected in differential amplification mode

where the new constant K' is related to K by the amplifier gain factor R_3/R_1. It is clear that a straight-line relationship now exists between the measurement signal y and the measured quantity x and that the unwanted bias has been removed.

5.5   Signal filtering

Signal filtering consists of processing a signal to remove a certain band of frequencies within it. The band of frequencies removed can be either at the low-frequency end of the frequency spectrum, at the high-frequency end, at both ends, or in the middle of the spectrum. Filters to perform each of these operations are known respectively as low-pass filters, high-pass filters, band-pass filters and band-stop filters. All such filtering operations can be carried out by either analog or digital methods.

The result of filtering can be readily understood if the analogy with a procedure such as sieving soil particles is considered. Suppose that a sample of soil A is passed through a system of two sieves of differing meshes such that the soil is divided into three parts, B, C and D, consisting of large, medium and small particles, as shown in figure 5.5. Suppose that the system also has a mechanism for delivering one or more of the separated parts, B, C and D, as the system output. If the graded soil output consists of parts C and D, the system is behaving as a low-pass filter (rejecting large particles), whereas if it consists of parts B and C, the system is behaving as a high-pass filter (rejecting small particles). Other options are to deliver just part C (band-pass filter mode) or parts B and D together (band-stop filter mode). As any gardener knows, however, such perfect seiving is not achieved in practice and any form of graded soil output always contains a few particles of the wrong size.

Signal filtering consists of selectively passing or rejecting low-, medium- and high-frequency signals from the frequency spectrum of a general signal. The range of frequencies passed by a filter is known as the *pass band*, the range not passed is

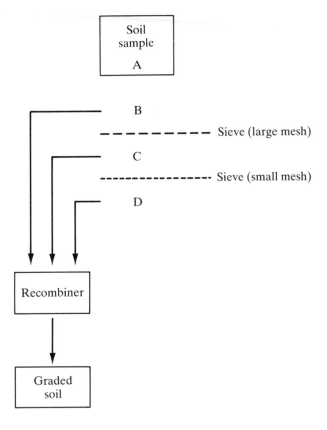

Figure 5.5   Soil sieving analogy of signal filtering

known as the *stop band*, and the boundary between the two ranges is known as the *cut-off frequency*. To illustrate this, consider a signal whose frequency spectrum is such that all frequency components in the frequency range from zero to infinity have equal magnitude. If this signal is applied to an ideal filter, then the outputs for a low-pass filter, high-pass filter, band-pass filter and band-stop filter respectively are shown in figure 5.6. Note that for the latter two types, the bands are defined by a pair of frequencies rather than by a single cut-off frequency.

Just as in the case of the soil sieving analogy presented above, the signal filter-ing mechanism is not perfect, with unwanted frequency components not being erased completely but only attenuated by varying degrees instead, i.e. the filtered signal always retains some components (of a relatively low magnitude) in the unwanted fre-quency range. There is also a small amount of attenuation of frequencies within the pass band which increases as the cut-off frequency is approached. Figure 5.7 shows the typical output characteristics of a practical constant-$k^\dagger$ filter designed respec-tively for high-pass, low-pass, band-pass and band-stop filtering. Filter design is con-cerned with trying to obtain frequency rejection characteristics which are as close to

---

† 'Constant-k' is a term used to describe a common class of passive filters, as discussed in the following section.

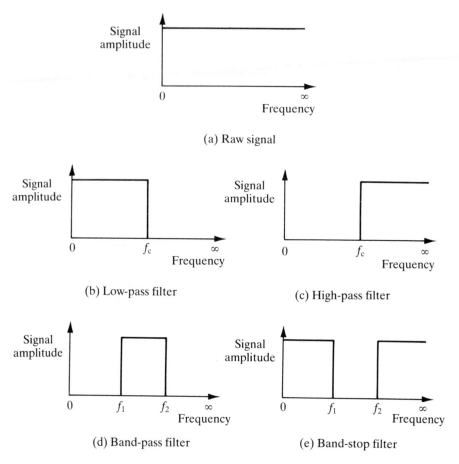

(a) Raw signal

(b) Low-pass filter

(c) High-pass filter

(d) Band-pass filter

(e) Band-stop filter

Figure 5.6   Outputs from ideal filters

the ideal as possible. However, improvement in characteristics is only achieved at the expense of greater complexity in the design. The filter chosen for any given situation is therefore a compromise between performance, complexity and cost.

In the majority of measurement situations, the physical quantity being measured has a value which is either constant or only changing slowly with time. In these circumstances, the most common types of signal corruption are high-frequency noise components, and the type of signal processing element required is a low-pass filter. In a few cases, the measured signal itself has a high frequency, for instance when mechanical vibrations are being monitored, and the signal processing required is the application of a high-pass filter to attenuate low-frequency noise components. Use of the other two types, band-stop and band-pass filters, is uncommon in measurement systems. These do however have some importance, with band-stop filters for instance having a potential application where a measurement signal is corrupted by noise at a particular frequency. Such noise is frequently due to mechanical vibrations or proximity of the measurement circuit to other electrical apparatus.

Both passive and active analog filter implementations of the four types of filter identified are considered below. The equivalent digital filters are discussed later in section 5.6.

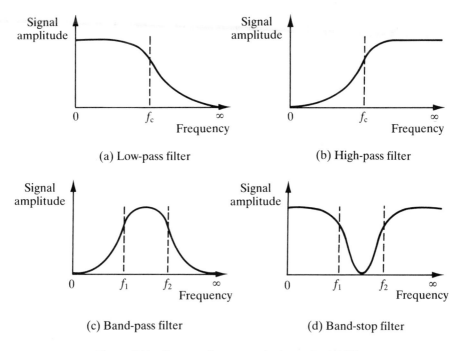

(a) Low-pass filter

(b) High-pass filter

(c) Band-pass filter

(d) Band-stop filter

Figure 5.7   Outputs from practical constant-*k* filters

Passive analog filters

The detailed design of passive filters is a subject of some considerable complexity which is outside the scope of this book. In the following treatment, the major formulas appropriate to the design of filters are quoted without derivation. The derivation can be found elsewhere (Blinchikoff (1976); Skilling (1967); Williams (1963)).

Simple passive filters consist of a network of impedances as shown in figure 5.8(a). So that there is no dissipation of energy in the filter, these impedances should ideally be pure reactances (capacitors or resistance-less inductors). In practice, however, it is impossible to manufacture inductors that do not have a small resistive component and so this ideal cannot be achieved. Indeed, readers familiar with radio receiver design will be aware of the existence of filters consisting of pure resistors and capacitors. These only have a mild filtering effect which is useful in radio tone controls but not relevant to the signal processing requirements discussed in this chapter. Such filters are therefore not considered further.

Each element of the network shown in figure 5.8(a) can be represented by either a *T*-section or π-section as shown in figures 5.8(b) and (c) respectively. To obtain proper matching between filter sections, it is necessary for the input impedance of each section to be equal to the load impedance for that section. This value of impedance is known as the characteristic impedance (Z_0). For a T-section of filter, the characteristic impedance is calculated from:

$$Z_0 = \{Z_1 Z_2 [1 + (Z_1/4Z_2)]\}^{1/2} \tag{5.8}$$

The frequency attenuation characteristics of the filter can be determined by inspect-

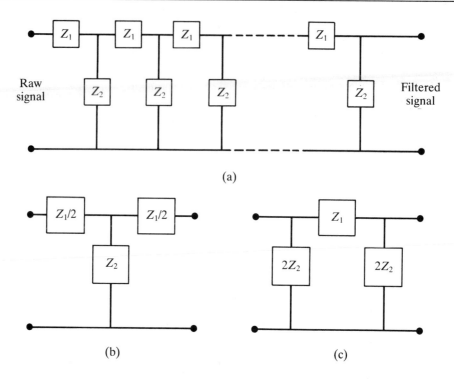

Figure 5.8 (a) Simple passive filter. (b) T-section. (c) π-section

ing this expression for Z_0. Frequency values for which Z_0 is real lie in the pass band and frequencies for which Z_0 is imaginary lie in the stop band.

Consider the case where $Z_1 = j\omega L$ and $Z_2 = 1/j\omega C$. Substituting these values into the expression for Z_0 above, we obtain:

$$Z_0 = [(L/C)(1 - 0.25\omega^2 LC)]^{1/2}$$

For frequencies where $\omega < (4/LC)^{1/2}$, Z_0 is real, and for higher frequencies, Z_0 is imaginary. These values of impedance therefore give a *low-pass filter* with cut-off frequency given by:

$$f_c = (\omega_c/2\pi) = \frac{1}{\pi(LC)^{1/2}}$$

A *high-pass filter* can be synthesized with exactly the same cut-off frequency if the impedance values chosen are:

$$Z_1 = 1/j\omega C \quad \text{and} \quad Z_2 = j\omega L$$

A point worthy of note in both these last two examples is that the product $Z_1 Z_2$ could be represented by a constant k which is independent of frequency. Because of this, such filters are known by the name of *constant-k filters*.

A constant-k *band-pass filter* can be realized with the following choice of impedance values:

$$Z_1 = j\omega L + \frac{1}{j\omega C} \qquad Z_2 = \frac{(j\omega La)(a/j\omega C)}{j\omega La + (a/j\omega C)}$$

The frequencies f_1 and f_2 defining the end of the pass band are most easily expressed in terms of a frequency f_0 in the center of the pass band. The corresponding equations are:

$$f_0 = \frac{1}{2\pi(LC)^{1/2}} \qquad f_1 = f_0[(1+a)^{1/2} - a^{1/2}] \qquad f_2 = f_0[(1+a)^{1/2} + a^{1/2}]$$

For a constant-k *band-stop filter*, the appropriate impedance values are:

$$Z_1 = \frac{(j\omega La)(a/j\omega C)}{j\omega La + (a/j\omega C)} \qquad Z_2 = \frac{1}{a}\left(j\omega L + \frac{1}{j\omega C}\right)$$

The frequencies defining the ends of the stop band are again normally defined in terms of the frequency f_0 in the center of the stop band:

$$f_0 = \frac{1}{2\pi(LC)^{1/2}} \qquad f_1 = f_0(1 - a/4) \qquad f_2 = f_0(1 + a/4)$$

The LC circuits forming these four types of filter T-section are summarized in figure 5.9.

As has already been mentioned, a practical filter does not eliminate frequencies in the stop band but merely attenuates them by a certain amount. The attenuation, α, at a frequency in the stop band, f, for a single T-section of a low-pass filter is given by:

$$\alpha = 2 \cosh^{-1}(f/f_c) \tag{5.9}$$

The relatively poor attenuation characteristics are obvious if we evaluate this expression for a value of frequency close to the cut-off frequency given by $f = 2f_c$. Then $\alpha = 2 \cosh^{-1}(2) = 2.64$. Further away from the cut-off frequency, for $f = 20f_c$, $\alpha = 2 \cosh^{-1}(20) = 7.38$.

Improved attenuation characteristics can be obtained by putting several T-sections in cascade. If perfect matching is assumed then two T-sections give twice the attenuation of one section, i.e. at frequencies of $2f_c$ and $20f_c$, α for two sections would have a value of 5.28 and 14.76 respectively.

The discussion so far has assumed resistance-less inductances and perfect matching between sections. Such conditions cannot be achieved in practice and this has several consequences.

Inspection of the expression for the characteristic impedance (5.8) reveals frequency-dependent terms. Thus the condition that the load impedance is equal to the input impedance for a section is only satisfied at one particular frequency. The

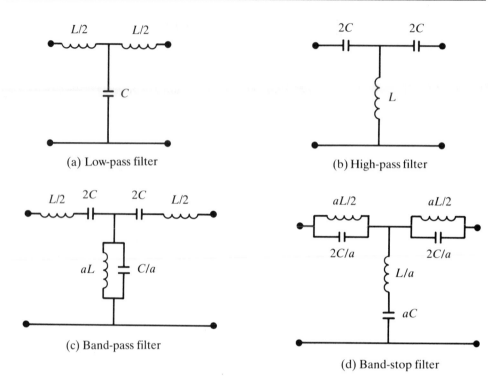

Figure 5.9   Circuit components for passive filter T-sections

load impedance is normally chosen so that this frequency is 'well within the pass band'. (Zero frequency is usually chosen for a low-pass filter and infinite frequency for a high-pass one as the frequency where this is satisfied.) This is one of the reasons for the degree of attenuation in the pass band shown in the practical filter characteristics of figure 5.7, the other reason being the presence of resistive components in the inductors of the filter. The effect of this in a practical filter is that the value of α at the cut-off frequency is 1.414 whereas the value predicted theoretically for an ideal filter (equation (5.9)) is 0. Cascading filter sections together increases this attenuation in the pass band as well as increasing attenuation of frequencies in the stop band.

This problem of matching successive sections in a cascaded filter seriously degrades the performance of constant-k filters and this has resulted in the development of other types such as m-derived and n-derived composite filters. These produce less attenuation within the pass band and greater attenuation outside it than constant-k filters, although this is only achieved at the expense of greater filter complexity and cost. The reader interested in further consideration of these is directed to one of the specialist texts recommended in the Further Reading section at the end of the book.

Active analog filters

In the foregoing discussion on passive filters, two main difficulties were encountered, of obtaining resistance-less inductors, and of achieving proper matching between signal source and load through the filter sections. Active filters overcome both of these problems and so are very popular for signal processing duties.

Active circuits to produce the four different types of filtering identified are illustrated in figure 5.10. These particular circuits are all second-order filters because the relationship between filter input and output is described by a second-order differential equation. The major component in an active filter is an electronic amplifier, with the filter characteristics being defined by a network of amplifier input and feedback components consisting of resistors and capacitors. The fact that resistors are used instead of the more expensive and bulky inductors required by passive filters is a further advantage of the active type of filter.

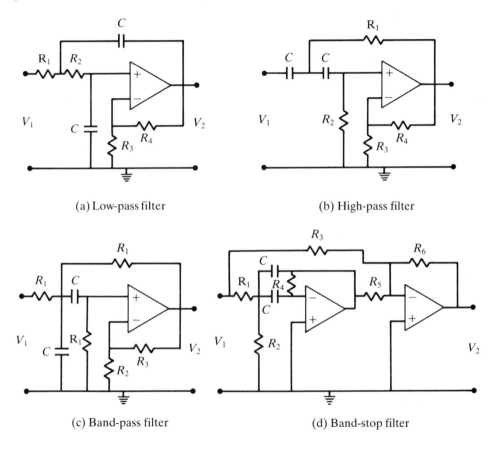

(a) Low-pass filter   (b) High-pass filter

(c) Band-pass filter   (d) Band-stop filter

Figure 5.10   Active filters

As in the case of passive filters, the design of active filters is a subject of considerable complexity and is only considered here in simple terms. The reader requiring a deeper understanding is referred to a specialist text, such as the one by Hilburn and Johnson (1973) mentioned in the Further Reading section.

The characteristics of the filter in terms of its attenuation behavior in the pass and stop bands is determined by the choice of circuit components in figure 5.10. A common set of design formulas are given below:

(i)   *Low-pass filter*

$$K = \frac{\mu}{R_1 R_2 C C_1}$$

$$a = \frac{1}{R_2 C_1}(1-\mu) + \frac{1}{R_1 C} + \frac{1}{R_2 C}$$

$$b = \frac{1}{R_1 R_2 C C_1}$$

where $\mu = 1 + R_4/R_3$ and d.c. gain is K/b.

(ii)   *High-pass filter*

$$K = \mu$$

$$a = \frac{1}{R_1 C}(1-\mu) + \frac{2}{R_2 C}$$

$$b = \frac{1}{R_1 R_2 C^2}$$

where $\mu = 1 + R_4/R_3$.

(iii)   *Band-pass filter*

$$K = \frac{\mu}{R_1 C}$$

$$B = \omega_2 - \omega_1 = \frac{4-\mu}{R_1 C}$$

$$\omega_0^2 = \frac{2}{R_1^2 C^2}$$

where $\mu = 1 + R_3/R_2$, ω_1 and ω_2 are frequencies at ends of the pass band, ω_0 is center frequency of pass band and gain at frequency ω_0 is K/b.

(iv)   *Band-stop filter*

$$R_3 R_4 = 2 R_1 R_5$$

$$B = \frac{2}{R_4 C}$$

$$\omega_0^2 = \frac{1}{R_4 C^2}\left(\frac{1}{R_1} + \frac{1}{R_2}\right)$$

where ω_1 and ω_2 are frequencies at ends of the stop band, ω_0 is center frequency of pass band and inverting gain magnitude is R_6/R_3.

Design procedures using the formulas above involve the use of graphs in which gains and cut-off frequencies are plotted for a range of parameter values (see Hilburn and Johnson 1973). Suitable parameter values are then chosen by inspection.

Whilst the above formulas represent the form of active filter which is the most general-purpose one available, many other forms also exist with the same circuit structure but with different rules for defining component values. Butterworth filters, for instance, optimize the pass-band attenuation characteristics at the expense of stop-band performance. Another form, Chebyshev filters, have very good stop-band attenuation characteristics but poorer pass-band performance. These are considered in considerable detail in the text by Hilburn and Johnson (1973).

5.6 Digital signal processing

Digital techniques achieve much greater levels of accuracy in signal processing than equivalent analog methods. However, the time taken to process a signal digitally is much longer than that required to carry out the same operation by analog techniques, and the equipment required is more expensive. Some care is needed therefore in making the correct choice between digital and analog methods in a particular signal processing application.

Whilst digital signal processing elements in a measurement system can exist as separate units, it is more usual to find them as an integral part of an intelligent instrument. However, their construction and mode of operation are the same irrespective of whether they exist physically as separate boxes or within an intelligent instrument.

The hardware aspect of a digital signal processing element consists of a digital computer and analog interface boards, whose general mode of operation has already been considered in chapter 3. The actual form which signal processing takes depends on the software program executed by the processor. Before consideration is given to this, however, some theoretical aspects of signal sampling need to be discussed.

As mentioned earlier, digital computers require signals to be in digital form whereas most instrumentation transducers have an output signal in analog form. Analog-to-digital conversion is therefore required at the interface between analog transducers and the digital computer. The procedure followed is to sample the analog signal at a particular moment in time and then convert the analog value to an equivalent digital one. This conversion takes a certain finite time, during which the analog signal can be changing in value. The next sample of the analog signal cannot be taken until the conversion of the last sample to digital form is completed. The representation within a digital computer of a continuous analog signal is therefore a sequence of samples whose pattern only approximately follows the shape of the original signal. This pattern of samples taken at successive, equal intervals of time is known as a discrete signal. The process of conversion between a continuous analog signal and a discrete digital one is illustrated for a sine wave in figure 5.11.

The raw analog signal in figure 5.11 has a frequency of approximately 0.75 cycles per second. With the rate of sampling shown, which is approximately 11

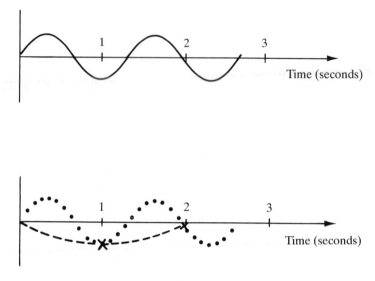

Figure 5.11   Conversion of continuous analog signal to discrete sampled signal

samples per second, reconstruction of the samples matches the original analog signal very well. If the rate of sampling were decreased, the fit between the reconstructed samples and the original signal would be less good. If the rate of sampling was very much less than the frequency of the raw analog signal, such as 1 sample per second, only the samples marked '×' in figure 5.11 would be obtained. Fitting a line through these '×'s incorrectly estimates a signal whose frequency is approximately 0.25 cycles per second. This phenomenon, whereby the process of sampling transmutes a high-frequency signal into a lower-frequency one, is known as *aliasing*. To avoid aliasing, it is necessary theoretically for the sampling rate to be at least twice the highest frequency in the analog signal sampled. In practice, sampling rates of between 5 and 10 times the highest frequency signal are normally chosen so that the discrete sampled signal is a close approximation to the original analog signal in amplitude as well as frequency.

Problems can arise in sampling when the raw analog signal is corrupted by high-frequency noise of unknown characteristics. It would be normal practice to choose the sampling interval as say a 10× multiple of the frequency of the measurement component in the raw signal. If such a sampling interval is chosen, aliasing can in certain circumstances transmute high-frequency noise components into the same frequency range as the measurement component in the signal, thus giving erroneous results. This is one of the circumstances mentioned earlier, where prior analog signal conditioning in the form of a low-pass filter must be carried out before processing the signal digitally.

One further factor which affects the quality of a signal when it is converted from analog to digital form is *quantization*. Quantization describes the procedure whereby the continuous analog signal is converted into a number of discrete levels. At any particular value of the analog signal, the digital representation is either the discrete level immediately above this value or the discrete level immediately below this value.

If the difference between two successive discrete levels is represented by the parameter Q, then the maximum error in each digital sample of the raw analog signal is $\pm Q/2$. This error is known as the quantization error and is clearly proportional to the resolution of the analog-to-digital converter, i.e. to the number of bits used to represent the samples in digital form.

Once a satisfactory digital representation in discrete form of an analog signal has been obtained, the procedures of signal amplification, signal attenuation and bias removal become trivial. For signal amplification and attenuation, all samples have to be multiplied or divided by a fixed constant. Bias removal involves simply adding or subtracting a fixed constant from each sample of the signal.

Signal linearization requires *a priori* knowledge of the type of non-linearity involved, in the form of a mathematical equation which expresses the relationship between the output measurements from an instrument and the value of the physical quantity being measured. This can be obtained either theoretically through knowledge of the physical laws governing the system or empirically using input–output data obtained from the measurement system under controlled conditions. Once this relationship has been obtained, it is used to calculate the value of the measured physical quantity corresponding to each discrete sample of the measurement signal. Whilst the amount of computation involved in this is greater than for the trivial cases of signal amplification, etc., already mentioned, the computational burden is still relatively small in most measurement situations.

Digital signal processing can also perform all of the filtering functions mentioned earlier in respect of analog filters, i.e. low-pass, high-pass, band-pass and band-stop. However, the design of digital filters requires a level of theoretical knowledge, including the use of z-transform theory, which is outside the range of this book. The reader interested in digital filter design is therefore referred elsewhere (Lynn (1973); Huelsman (1970)).

6

Measurement System Errors

It is extremely important in any measurement system to reduce errors in instrument output readings to the minimum possible level and to quantify the maximum error which may exist in any output reading. This requires a detailed analysis of the sources of error which exist in a measurement system as a prerequisite in this process of reducing the total measurement error level. Errors in measurement data can be divided into two groups, known as random errors and systematic errors. This chapter is concerned with identifying the various errors which exist under these headings, and suggesting mechanisms for reducing their magnitude and effect. The way in which these separate error components are assessed in order to calculate the overall measurement system error level is considered later in chapter 7.

6.1 Random and systematic errors

Random errors usually arise when measurements are taken by human observation of an analog meter reading, especially where this involves interpolation between scale points. Electrical noise can also be a source of random errors. To some extent, random errors can be overcome by taking the same measurement a number of times and extracting a value by statistical means, as discussed in chapter 7. At each attempt, a human observer will make a different observation, but the mean of these readings is likely to be close to the correct value as long as the human observer is not introducing a systematic parallax-induced error by persistently reading an analog meter from the same side rather than from directly above.

Systematic errors describe errors in the output readings of a measurement system which are unlikely to be revealed by repeated readings, and these therefore cause greater problems. Two major sources of systematic errors are system disturbance during measurement and the effect of modifying inputs, as discussed in sections 6.2 and 6.3. Other sources of systematic error include bent meter needles, the use of uncalibrated instruments, cabling practices and thermal e.m.f.'s. The latter two sources are considered in section 6.4. Even when systematic errors due to the above factors have been reduced or eliminated, some errors remain which are inherent in the manufacture of an instrument. These are quantified in the accuracy figure quoted in the published instrument specifications.

6.2   System disturbance due to measurement

If we were to start with a beaker of hot water and wished to measure its temperature with a mercury-in-glass thermometer, then we should take the thermometer, which would be initially at room temperature, and plunge it into the water. In so doing, we would be introducing a relatively cold mass (the thermometer) into the hot water and a heat transfer would take place between the water and the thermometer. This heat transfer would lower the temperature of the water. Whilst in this case the reduction in temperature would be so small as to be undetectable by the limited measurement

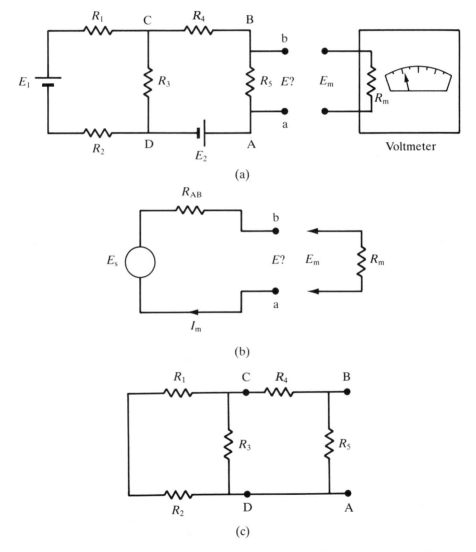

Figure 6.1 (a) A circuit in which the voltage across R_5 is to be measured.
(b) Equivalent circuit by Thevenin's theorem.
(c) The circuit used to find the equivalent single resistance R_{AB}

resolution of such a thermometer, the effect is finite and clearly establishes the principle that, in nearly all measurement situations, the process of measurement disturbs the system and alters the value of the physical quantities being measured.

Another example is that of measuring car tire pressures with the type of pressure gauge commonly obtainable from car accessory shops. Measurement is made by pushing one end of the pressure gauge onto the valve of the tire and reading the displacement of the other end of the gauge against a scale. As the gauge is used, a quantity of air flows from the tire into the gauge. This air does not subsequently flow back into the tire after measurement, and so the tire has been disturbed and the air pressure inside it has been permanently reduced.

Thus, as a general rule, the process of measurement always disturbs the system being measured. The magnitude of the disturbance varies from one measurement system to the next and is affected particularly by the type of instrument used for measurement. Ways of minimizing disturbance of measured systems is an important consideration in instrument design. A prerequisite for this, however, is an accurate understanding of the mechanisms of system disturbance.

Measurements in electric circuits

In analyzing system disturbance during measurements in electric circuits, Thevenin's theorem (see appendix 3) is often of great assistance. Consider for instance the circuit shown in figure 6.1(a) in which the voltage across resistor R_5 is to be measured by a voltmeter with resistance R_m. Here R_m acts as a shunt resistance across R_5, decreasing the resistance between points AB and so disturbing the circuit. The voltage E_m measured by the meter is not therefore the value of the voltage E_0 that existed prior to measurement. The extent of the disturbance can be assessed by calculating the open-circuit voltage E_0 and comparing it with E_m.

Thevenin's theorem allows the circuit of figure 6.1(a) comprising two voltage sources and five resistors to be replaced by an equivalent circuit containing a single resistance and one voltage source, as shown in figure 6.1(b). For the purpose of defining the equivalent single resistance of a circuit by Thevenin's theorem, all voltage sources are represented just by their internal resistance, which can be approximated to zero, as shown in figure 6.1(c). Analysis proceeds by calculating the equivalent resistances of sections of the circuit and building these up until the required equivalent resistance of the whole of the circuit is obtained. So, in (c), starting at C and D, the circuit to the left of C and D consists of a series pair of resistances (R_1 and R_2) in parallel with R_3, and the equivalent resistance can be written as:

$$\frac{1}{R_{CD}} = \frac{1}{R_1 + R_2} + \frac{1}{R_3} \quad \text{or} \quad R_{CD} = \frac{(R_1 + R_2)R_3}{R_1 + R_2 + R_3}$$

Moving now to A and B, the circuit to the left consists of a pair of series resistances (R_{CD} and R_4) in parallel with R_5. The equivalent circuit resistance R_{AB} can thus be written as:

$$\frac{1}{R_{AB}} = \frac{1}{R_{CD} + R_4} + \frac{1}{R_5}$$

or

$$R_{AB} = \frac{(R_4 + R_{CD})R_5}{R_4 + R_{CD} + R_5}$$

Substituting for R_{CD} using the previously derived expression, we obtain:

$$R_{AB} = \frac{[(R_1 + R_2)R_3/(R_1 + R_2 + R_3) + R_4]R_5}{(R_1 + R_2)R_3/(R_1 + R_2 + R_3) + R_4 + R_5} \qquad (6.1)$$

Defining I as the current flowing in the circuit when the measuring instrument is connected to it, we can write:

$$I = \frac{E_0}{R_{AB} + R_m}$$

and the voltage measured by the meter is given by:

$$E_m = \frac{R_m E_0}{R_{AB} + R_m}$$

In the absence of the measuring instrument and its resistance R_m, the voltage across AB would be the equivalent circuit voltage source whose value is E_0. The effect of measurement is therefore to reduce the voltage across AB by the ratio given by:

$$\frac{E_m}{E_0} = \frac{R_m}{R_{AB} + R_m} \qquad (6.2)$$

It is thus obvious that as R_m gets larger, the ratio E_m/E_0 gets closer to unity, showing that the design strategy should be to make R_m as high as possible to minimize disturbance of the measured system. (Note that we did not calculate the value of E_0, since this is not required in quantifying the effect of R_m.)

Example 1

Suppose that the components of the circuit shown in figure 6.1(a) have the following values: $R_1 = 400\ \Omega$, $R_2 = 600\ \Omega$, $R_3 = 1000\ \Omega$, $R_4 = 500\ \Omega$, $R_5 = 1000\ \Omega$. The voltage across AB is measured by a voltmeter whose internal resistance is $9500\ \Omega$. What is the measurement error caused by the resistance of the measuring instrument?

Solution

Proceeding by applying Thevenin's theorem to find an equivalent circuit to that of figure 6.1(a) of the form shown in figure 6.1(b), and substituting the given component values into the equation for R_{AB} (6.1), we obtain:

$$R_{AB} = \frac{(1000^2/2000 + 500)\,1000}{1000^2/2000 + 500 + 1000} = \frac{1000^2}{2000} = 500\,\Omega$$

From equation (6.2), we have:

$$\frac{E_m}{E_0} = \frac{R_m}{R_{AB} + R_m}$$

The measurement error is given by $(E_0 - E_m)$

$$E_0 - E_m = E_0 \left(1 - \frac{R_m}{R_{AB} + R_m}\right)$$

Substituting in values:

$$E_0 - E_m = E_0 \left(1 - \frac{9500}{10000}\right) = 0.95 E_0$$

Thus the error in the measured value is 5%.

At this point, it is interesting to note what constraints exist when practical attempts are made to achieve a high internal resistance in the design of a moving-coil voltmeter. Such an instrument consists of a coil carrying a pointer mounted in a fixed magnetic field. As current flows through the coil, the interaction between the field generated and the fixed field causes the pointer it carries to turn in proportion to the applied current (see chapter 8, especially figure 8.2, for further details).

The simplest way of increasing the input impedance (the resistance) of the meter is either to increase the number of turns in the coil or to construct the same number of coil turns with a higher-resistance material. Either of these solutions decreases the current flowing in the coil, however, giving less magnetic torque and thus decreasing the measurement sensitivity of the instrument (i.e. for a given applied voltage, we get less deflection of the pointer). This problem can be overcome by changing the spring constant of the restraining springs of the instrument, such that less torque is required to turn the pointer by a given amount. This however reduces the ruggedness of the instrument and also demands better pivot design to reduce friction. This highlights a very important but tiresome principle in instrument design: any attempt to improve the performance of an instrument in one aspect generally decreases the performance in some other aspect. This is an inescapable fact of life with passive instruments such as the type of voltmeter mentioned, and is often the reason for the use of alternative active instruments such as digital voltmeters, where the inclusion of auxiliary power greatly improves performance.

The use of deflection-type bridge circuits is another area where careful thought has to be given to the design of the instrument measuring the output voltage if the process of measurement is not to interfere with the value of the measured quantity. In the analysis of bridge circuits in the last chapter, a simplification was made by assuming zero current in the measuring instrument. As at least a small, finite current must flow into the measuring instrument, the zero-current assumption clearly intro-

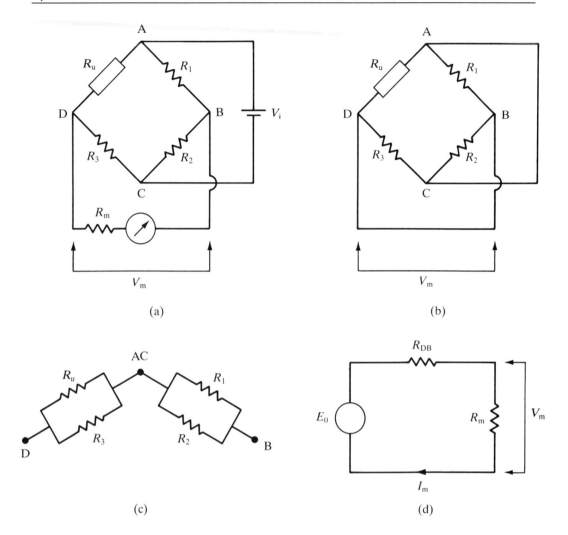

Figure 6.2 (a) A bridge circuit. (b) Equivalent circuit by Thevenin's theorem.
(c) Alternate representation. (d) Equivalent circuit via Thevenin's theorem

duces some error which must be quantified. Thevenin's theorem is again a useful tool
for this purpose. Replacing the voltage source V_i in figure 6.2(a) by a zero internal
resistance produces the circuit shown in figure 6.2(b) which can be shown in an
equivalent representation (figure 6.2(c)). Figure 6.2(c) shows that the equivalent
circuit resistance consists of a pair of parallel resistors R_u and R_3 in series with the
parallel resistor pair R_1 and R_2, and R_{DB} is given by:

$$R_{DB} = \frac{R_1 R_2}{R_1 + R_2} + \frac{R_u R_3}{R_u + R_3} \tag{6.3}$$

The equivalent circuit derived via Thevenin's theorem with the resistance R_m of the
measuring instrument connected across the output is shown in figure 6.2(d).

The open-circuit voltage across DB, E_0, is the output voltage calculated in chapter 4 (equation (4.3)) for the case of $R_m = 0$:

$$E_0 = V_i \left(\frac{R_u}{R_u + R_3} - \frac{R_1}{R_1 + R_2} \right) \qquad (6.4)$$

If the current flowing is I_m when the measuring instrument of resistance R_m is connected across DB, then, by Ohm's law, I_m is given by:

$$I_m = \frac{E_0}{R_{DB} + R_m} \qquad (6.5)$$

If V_m is the voltage measured across R_m, then, again by Ohm's law:

$$V_m = I_m R_m = \frac{E_0 R_m}{R_{DB} + R_m} \qquad (6.6)$$

Substituting for E_0 and R_{DB} in equation (6.6), using the relationships developed in equations (6.3) and (6.4), we obtain:

$$V_m = \frac{V_i [R_u/(R_u + R_3) - R_1/(R_1 + R_2)] R_m}{R_1 R_2/(R_1 + R_2) + R_u R_3/(R_u + R_3) + R_m}$$

Simplifying:

$$V_m = \frac{V_i R_m (R_u R_2 - R_1 R_3)}{R_1 R_2 (R_u + R_3) + R_u R_3 (R_1 + R_2) + R_m (R_1 + R_2)(R_u + R_3)} \qquad (6.7)$$

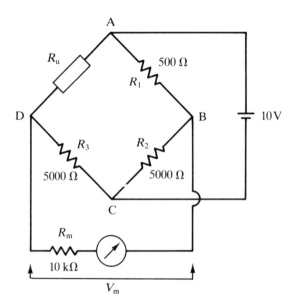

Figure 6.3   Bridge circuit (example 2)

Example 2

A bridge circuit as shown in figure 6.3 is used to measure the value of the unknown resistance R_u of a strain gauge of nominal value 500 Ω. The output voltage measured across points DB in the bridge is measured by a voltmeter. Calculate the measurement sensitivity in volts per ohm change in R_u if (a) the resistance R_m of the measuring instrument is neglected, and (b) account is taken of the value of R_m.

Solution

For $R_u = 500\,\Omega$, $V_m = 0$. To determine sensitivity, calculate V_m for $R_u = 501\,\Omega$.
(a) Applying equation (4.3):

$$V_m = V_i \left(\frac{R_u}{R_u + R_3} - \frac{R_1}{R_1 + R_2} \right)$$

Substituting in values:

$$V_m = 10 \left(\frac{501}{1001} - \frac{500}{1000} \right) = 5.00\,\text{mV}$$

Thus, if the resistance of the measuring circuit is neglected, the measurement sensitivity is 5.00 mV per ohm change in R_u. (b) Applying equation (6.5) and substituting in values:

$$V_m = \frac{10 \times 10^4 \times 500 \times (501-500)}{500^2 \times (1001) + 500 \times 501 \times (1000) + 10^4 \times 1000 \times 1001} = 4.76\,\text{mV}$$

Thus, if account is taken of the 10 kΩ value of the resistance of R_m, the measurement sensitivity is 4.76 mV per ohm change in R_u.

6.3 Modifying inputs in measurement systems

The fact that the static and dynamic characteristics of measuring instruments are specified for particular environmental conditions of temperature and pressure, etc., has already been discussed at considerable length in chapter 2. These specified conditions must be reproduced as closely as possible during calibration exercises because, away from the specified calibration conditions, the characteristics of measuring instruments vary to some extent. The magnitude of this variation is quantified by the two constants known as the sensitivity drift and the zero drift, both of which are generally included in the published specifications for an instrument. Such variations of environmental conditions away from the calibration conditions are described as modifying inputs to the system. The environmental variation is described as a measurement system input because the effect on the system output is the same as if the value of the measured quantity had changed by a certain amount. In any general measurement situation, it is very difficult to avoid modifying inputs, because it is either impractical or impossible to control the environmental conditions surrounding the measurement system. System designers are therefore charged with the task of either reducing the susceptibility of measuring instruments to modifying inputs or alternately quantifying the effect of modifying inputs and correcting for them in the instrument output reading.

The techniques used to deal with modifying inputs and minimize their effect on the final output measurement follow a number of routes as discussed below.

Careful instrument design

Careful instrument design is the most useful weapon in the battle against modifying inputs, by reducing the sensitivity of an instrument to modifying inputs to as low a level as possible. In the design of strain gauges for instance, the element should be constructed from a material whose resistance has a very low temperature coefficient (i.e. the variation of the resistance with temperature is very small). For many instruments, however, it is not possible to reduce their sensitivity to modifying inputs to a satisfactory level by simple design adjustments, and other measures have to be taken.

Method of opposing inputs

The method of opposing inputs compensates for the effect of a modifying input in a

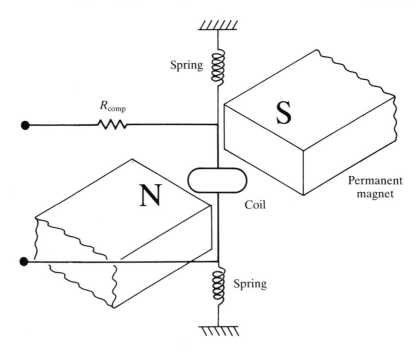

Figure 6.4   Millivoltmeter

measurement system by introducing an equal and opposite modifying input which cancels it out. One example of how this technique is applied is in the type of millivolt-meter shown in figure 6.4. This consists of a coil suspended in a fixed magnetic field produced by a permanent magnet. When an unknown voltage is applied to the coil, the magnetic field due to the current interacts with the fixed field and causes the coil (and a pointer attached to the coil) to turn. If the coil resistance is sensitive to temperature, then any modifying input to the system in the form of a temperature change will alter the value of the coil current for a given applied voltage and so alter the pointer output reading. Compensation for this is made by introducing a compensating resistance R_{comp} into the circuit, where R_{comp} has a temperature coefficient which is equal in magnitude but opposite in sign to that of the coil.

High-gain feedback

The benefit of adding high-gain feedback to many measurement systems is illustrated by considering the case of the voltage-measuring instrument whose block diagram is shown in figure 6.5. In this system, the unknown voltage E_i is applied to a motor of torque constant K_m, and the torque induced turns a pointer against the restraining action of a spring with spring constant K_s. The effect of modifying inputs on the motor and spring constants are represented by variables D_m and D_s.

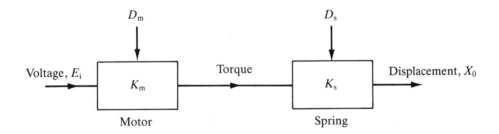

Figure 6.5   Block diagram of voltage-measuring instrument

The displacement of the pointer X_0 is given by:

$$X_0 = K_m K_s E_i$$

in the absence of modifying inputs. In the presence of modifying inputs, however, both K_m and K_s change and the relationship between X_0 and E_i can be affected greatly. It therefore becomes difficult or impossible to calculate E_i from the measured value of X_0.

Consider now what happens if the system is converted into a high-gain, closed-loop one, as shown in figure 6.6, by adding an amplifier of gain constant K_a and a feedback device with gain constant K_f. Assume also that the effect of modifying inputs on the values of K_a and K_f are represented by D_a and D_f. The feedback device feeds back a voltage E_0 proportional to the pointer displacement X_0. This is compared with the unknown voltage E_i by a comparator and the error is amplified.

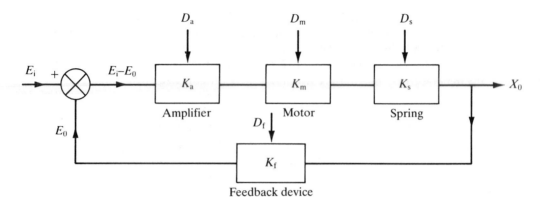

Figure 6.6   Conversion of system of figure 6.5 into a high-gain, closed-loop one

Writing down the equations of the system, we have:

$$E_0 = K_f X_0$$

$$X_0 = (E_i - E_0)K_a K_m K_s = (E_i - K_f X_0)K_a K_m K_s$$

Thus:

$$E_i K_a K_m K_s = (1 + K_f K_a K_m K_s)X_0$$

i.e.

$$X_0 = \frac{K_a K_m K_s}{1 + K_f K_a K_m K_s} E_i \qquad (6.8)$$

Because K_a is very large (it is a high-gain amplifier), $K_f K_a K_m K_s \gg 1$, and equation (6.8) reduced to:

$$X_0 = E_i/K_f$$

This is a highly important result because we have reduced the relationship between X_0 and E_i to one which involves only K_f. The sensitivity of the gain constants K_a, K_m and K_s to the modifying inputs D_a, D_m and D_s has thereby been rendered irrelevant and we only have to be concerned with one modifying input D_f. Conveniently, it is usually an easy matter to design a feedback device which is insensitive to modifying inputs: this is much easier than trying to make a motor or spring insensitive. Thus high-gain feedback techniques are often a very effective way of reducing a measurement system's sensitivity to modifying inputs. One potential problem which must be mentioned however is that there is a possibility that high-gain feedback will cause instability in the system. Any application of this method must therefore include careful stability analysis of the system.

Signal filtering

One frequent problem in measurement systems is corruption of the output reading by periodic noise, often at a frequency of 50 Hz caused by pick-up through the close proximity of the measurement system to apparatus or current-carrying cables operating on a mains supply. Periodic noise corruption at higher frequencies is also often introduced by mechanical oscillation or vibration within some component of a measurement system. The amplitude of all such noise components can be substantially attenuated by the inclusion of filtering of an appropriate form in the system, as discussed at great length in the last chapter. Band-stop filters can be especially useful where corruption is of one particular known frequency, or, more generally, low-pass filters are employed to attenuate all noise in the range of 50 Hz and higher frequencies.

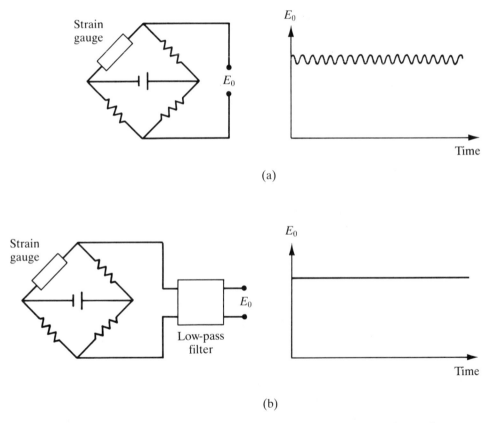

Figure 6.7 (a) Noise-corrupted output of bridge circuit measuring resistance of strain gauge. (b) Effect of adding low-pass filter

Measurement systems with a low-level output, such as a bridge circuit measuring a strain-gauge resistance, are particularly prone to noise, and figure 6.7(a) shows the typical corruption of a bridge output by 50 Hz pick-up. The beneficial effect of putting a simple passive RC low-pass filter across the output is shown in figure 6.7(b).

6.4   Other sources of error

Connecting leads

In connecting together the components of a measurement system, a common source of error is the failure to take proper account of the resistance of connecting leads (or pipes in the case of pneumatically or hydraulically actuated measurement systems). In typical applications of a resistance thermometer for instance, it is common to find the thermometer separated from other parts of the measurement system by perhaps 30 m. The resistance of such a length of 7/0.0076 copper wire is 2.5 Ω and there is a further complication that such wire has a temperature coefficient of 1 mΩ/°C.

Careful consideration therefore needs to be given to the choice of connecting leads. Not only should they be of adequate cross-section so that their resistance is minimized, but they should be adequately screened if they are thought likely to be subject to electrical or magnetic fields which could otherwise cause induced noise. Where screening is thought essential, then the routing of cables also needs careful planning. In one application in the author's personal experience involving instrumentation of an electric-arc steelmaking furnace, screened signal-carrying cables between transducers on the arc furnace and a control room at the side of the furnace were initially corrupted by high-amplitude 50 Hz noise. However, by changing the route of the cables between the transducers and the control room, the magnitude of this induced noise was reduced by a factor of about 10.

Thermal e.m.f.'s

Whenever metals of two different types are connected together, a thermal e.m.f. is generated according to the temperature of the joint. This is the physical principle on which temperature-measuring thermocouples operate (see chapter 15). Such thermal e.m.f.'s are only of a few millivolts in magnitude and so this effect is only significant when typical voltage output signals of a measurement system are of a similar low magnitude.

One such situation is where one e.m.f.-measuring instrument is used to monitor the output of several thermocouples measuring the temperatures at different points in a process control system. This requires a means of automatically switching the output of each thermocouple to the measuring instrument in turn. Nickel–iron reed relays with copper connecting leads are commonly used to provide this switching function. This introduces a thermocouple effect of magnitude 40 μV/°C between the reed relay and the copper connecting leads. There is no problem if both ends of the reed relay are at the same temperature because then the thermal e.m.f.'s will be equal and opposite and so cancel out. There are several recorded instances however where, because of lack of awareness of the problem, poor design has resulted in the two ends of a reed relay being at different temperatures and resulting in a net thermal e.m.f. The serious error that this introduces is clear. For a temperature difference between the two ends of only 2°C, the thermal e.m.f. is 80 μV, which is very large compared with a typical thermocouple output level of 400 μV.

Another example of the problems thermal e.m.f.'s can create becomes apparent in considering the following current-measuring system, where the current flowing is calculated by applying it to an accurately calibrated wire-wound resistance of value 100 Ω and measuring the voltage drop across the resistance. In calibration of the system, a known current of 20 μA was applied to the resistance and a voltage of 2.20 mV was measured by an accurate high-impedance instrument. Simple application of Ohm's law reveals that such a voltage reading indicates a current value of 22 μA. What then is the explanation for this discrepancy? The answer once again is a thermal e.m.f. Because the designer was not aware of thermal e.m.f.'s, the circuit had been constructed such that one side of the standard resistance was close to a power transistor, creating a difference in temperature between the two ends of the resistor of 2°C. The thermal e.m.f. associated with this was sufficient to account for the 10% measurement error found.

6.5   Error reduction using intelligent instruments

Intelligence in instruments can bring about a gross reduction in the level of output error in measurement systems which are subject to the types of error discussed earlier in this chapter. The ability of intelligent instruments to achieve this does however require that the following preconditions be satisfied:

(i)   The physical mechanism by which a measurement transducer is affected by ambient condition changes must be fully understood and all physical quantities which affect the transducer output must be identified.

(ii)   The effect of each ambient variable on the output characteristic of the measurement transducer must be quantified.

(iii)   Suitable secondary transducers for monitoring the value of all relevant ambient variables must be available for input to the intelligent instrument.

Condition (i) above means that the thermal expansion/contraction of all elements within a transducer must be considered in order to evaluate how it will respond to ambient temperature changes. Similarly, the transducer response, if any, to changes in ambient pressure, humidity, gravitational force or power supply level (active instruments) must be examined.

Quantification of the effect of each ambient variable on the characteristics of the measurement transducer is then necessary, as stated in condition (ii). Analytic quantification of ambient condition changes from purely theoretical consideration of the construction of a transducer is usually extremely complex and so is normally avoided. Instead, the effect is quantified empirically in laboratory tests where the output characteristic of the transducer is observed as the ambient environmental conditions are changed in a controlled manner.

Once the ambient variables affecting a measurement transducer have been identified and their effect quantified, an intelligent instrument can be designed which includes secondary transducers to monitor the value of the ambient variables. Suitable transducers which will operate satisfactorily within the environmental conditions prevailing for the measurement situation must of course exist, as stated in condition (iii).

Application of intelligent instruments to reduce random errors

Most cases of random error in measurement systems arise because of human errors in observing meters, and intelligent instruments can make no contribution toward reducing these. Some sources of random error arise however from the presence of electrical noise or mechanical vibrations which affect the measurement system. In such cases, intelligent instruments are programmed to take the same measurement a number of times within a short space of time and perform simple averaging on the readings before displaying an output measurement.

Application of intelligent instruments to reduce systematic errors

The contribution of intelligent instruments in the reduction of systematic errors is subject to the three conditions already stated, and the degree of applicability varies according to the features of each particular measurement situation. Each measurement situation must therefore be considered separately.

In the case of electrical circuits in which the system is disturbed by the loading effect of the measuring instrument, an intelligent instrument can readily correct for measurement errors by applying equations such as (6.2) with the resistance of the measuring instrument inserted. Similarly for measurement situations containing bridge circuits, equation (6.7) can be manipulated into a form where R_u is expressed as a function of V_i, V_m and the other circuit resistances if the resistance of the measuring instrument R_m is known. Then an intelligent instrument can use that expression to compute the unknown value of R_u.

Intelligent instruments are particularly useful for dealing with measurement errors due to modifying inputs. Secondary transducers are provided within the instrument to monitor the magnitude of environmental conditions such as temperature and pressure which can affect the characteristics of the primary measurement transducer. The computer within the instrument then corrects the measurement obtained from the primary transducer according to the values read by the secondary transducers. This presupposes that all the modifying inputs in a measurement situation have been correctly identified and quantified, and that suitable secondary transducers exist to monitor these modifying inputs.

Suitable care must always be taken when introducing a microcomputer into a measurement system to avoid creating new sources of measurement noise. This is particularly so where one microcomputer is used to process the output of several transducers and is connected to them by signal wires. In such circumstances, the connections and connecting wires can create noise through electrochemical potentials, thermoelectric potentials, offset voltages introduced by common mode impedances, and a.c. noise at power, audio and radio frequencies. Recognition of all these possible noise sources allows them to be eliminated in most cases by employing good measurement system construction practices. All remaining noise sources are usually eliminated by the provision of a set of four earthing circuits within the interface fulfilling the following functions:

(i) *Power earth:* provides a path for fault currents due to power faults.
(ii) *Logic earth:* provides a common line for all logic circuit potentials.
(iii) *Analog earth* (ground): provides a common reference for all analog signals.

(iv) *Safety earth:* connected to all metal parts of equipment to protect personnel should power lines come into contact with metal enclosures.

Exercises

6.1 Instruments are normally calibrated and their characteristics defined for particular standard ambient conditions. What procedures are normally taken to avoid measurement errors when using instruments which are subjected to changing ambient conditions?

6.2 The voltage across a resistance R_5 in the circuit of figure 6.8 is to be measured by a voltmeter connected across it.

Figure 6.8   Circuit for exercise 6.2

(a) If the voltmeter has an internal resistance (R_m) of 4750 Ω, what is the measurement error?
(b) What value would the voltmeter internal resistance need to be in order to reduce the measurement error to 1%?

6.3 In the circuit shown in figure 6.9, the current flowing between A and B is measured by an ammeter whose internal resistance is 100 Ω. What is the measurement error caused by the resistance of the measuring instrument?

Figure 6.9   Circuit for exercise 6.3

6.4   The output of a potentiometer is measured by a voltmeter having a resistance R_m, as shown in figure 6.10. R_t is the resistance of the total length X_t of the potentiometer and R_i is the resistance between the wiper and common point C for a general wiper position X_i. Show that the measurement error due to the resistance R_m of the measuring instrument is given by:

$$\text{Error} = E\left(\frac{R_i^2(R_t - R_i)}{R_t(R_iR_t + R_mR_t - R_i^2)} \right)$$

Hence show that the maximum error occurs when X_i is approximately equal to $2X_t/3$. [*Hint* – Differentiate the error expression with respect to R_i and set to 0. Note that maximum error does not occur exactly at $X_i = 2X_t/3$, but this value is very close to the position where the maximum error occurs.]

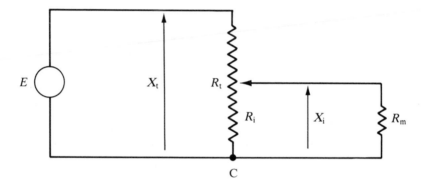

Figure 6.10   Circuit for exercise 6.4

7

Data Analysis and Presentation

The various sources of error which might occur in a measurement system, and the measures which can be taken to overcome them, have been discussed at length in chapter 6. When measurement errors have been reduced as far as possible, the remaining errors must be quantified and their total effect on measurement system accuracy calculated.

This chapter is concerned with the analysis of a set of data forming the output of a measurement system in order to extract the maximum amount of information from it and to present the data in the most useful form. Quantification of each separate systematic and random error is required first, following which the cumulative effect of all errors on overall measurement system accuracy can be calculated, as indicated later in section 7.2.

Systematic errors arising from tolerances in system components and other systematic errors inherent in the manufacture of instruments are quantified by published specifications for components and instruments, in the form of quoted accuracy and tolerance values, as discussed in chapter 1. The error due to other systematic error sources can be assessed as indicated in chapter 6. The remaining sources of error are random errors, for which statistical analysis is required, as described in section 7.1.

7.1 Statistical analysis of data

Mean and median values

In any measurement situation subject to random errors, the normal technique is to take the same reading a number of times, ideally using different observers, and extract the most likely value from the measurement data set. For a set of n measurements $x_1, x_2, \ldots , x_n$, the most likely true value is the *mean* given by:

$$x_{\text{mean}} = \frac{x_1 + x_2 + \ldots + x_n}{n} \tag{7.1}$$

This is valid for all data sets where the measurement errors are distributed equally about the line of zero error, i.e. where the positive errors are balanced in quantity and magnitude by the negative errors.

When the number of values in the data set is large however, calculation of the mean value is tedious, and it is more convenient to use the median value, this being a close approximation to the mean value. The *median* is given by the middle value when the measurements in the data set are written down in ascending order of magnitude.

For a set of n measurements $x_1, x_2, \ldots, x_n$ written down in ascending order of magnitude, the median value is given by:

$$x_{median} = x_{(n+1)/2}$$

Thus, for a set of nine measurements $x_1, x_2, \ldots, x_9$, the median value is x_5. For an even number of data values, the median value is midway between the center two values, i.e. for 10 measurements $x_1, \ldots, x_{10}$, the median value is given by $(x_5 + x_6)/2$.

Suppose that, in a particular measurement situation, a mass is measured by a beam balance, and the following set of readings is obtained at a particular time by different observers.

$$\begin{array}{cccccccccc} 81.6 & 81.1 & 81.4 & 80.9 & 81.1 & 80.5 & 81.3 & 80.8 & 81.2 & 81.8 \\ 81.1 & 81.5 & 81.0 & 81.3 & 81.1 & 80.8 & 81.3 & 81.6 & 81.1 \end{array} \quad (7.2)$$

The mean value of this set of data is 81.18, calculated according to equation (7.1). The median value is 81.1, which is the middle value if the data values are written down in ascending order, starting at 80.5 and ending at 81.8.

Standard deviation and variance

The probability that the mean or median value of a data set represents the true measurement value depends on how widely scattered the data values are. If the values in the last set of mass measurements (7.2) had ranged from 79 up to 83, our confidence in the mean value would be much less. The spread of values about the mean is analyzed by first calculating the deviation of each value from the mean. For any general value x_i, the deviation d_i is given by:

$$d_i = x_i - x_{mean}$$

The extent to which n measurement values are spread about the mean can now be expressed by the standard deviation σ, where σ is given by:

$$\sigma = \left(\frac{d_1^2 + d_2^2 + \ldots + d_n^2}{n-1} \right)^{1/2} \quad (7.3)$$

This spread can alternately be expressed by the variance V, which is the square of the standard deviation, i.e.

$$V = \sigma^2$$

Frequency distributions

A further and very powerful way of analyzing the pattern in which measurements deviate from the mean value is to use graphical techniques. The simplest way of doing this is by means of a *histogram*, where various error bands of equal widths are defined and the number of measurements with error within each range recorded. Figure 7.1 shows a histogram drawn from the set of mass measurement data in (7.2). This has the characteristic shape shown by truly random data, with symmetry about the zero error line and steps of increasing height on either side going up to the peak.

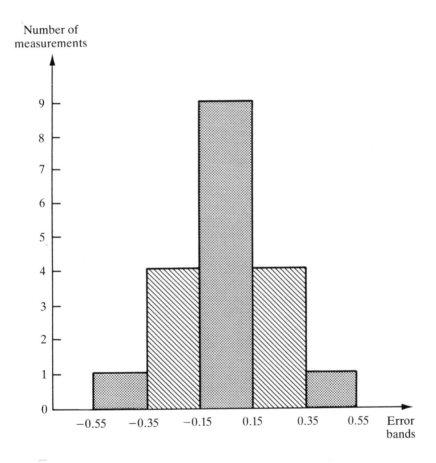

Figure 7.1   Histogram of errors

As the number of measurements increases, smaller error bands can be defined for the error histogram, which retains its basic shape but then consists of a larger number of smaller steps on each side of the peak. In the limit, as the number of measurements approaches infinity, the histogram becomes a smooth curve known as a *frequency distribution curve* as shown in figure 7.2. The ordinate of this curve is the frequency of occurrence of each error level, $f(E)$, and the abscissa is the error magnitude, E.

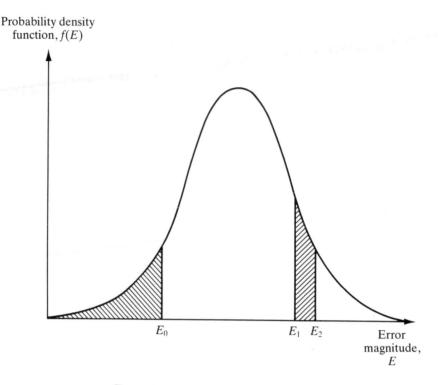

Figure 7.2   Frequency distribution of errors

If the height of the frequency distribution curve is standardized such that the area under it is unity, then the curve in this form is known as a *probability curve*, and the height $f(E)$ at any particular error magnitude E is known as the *probability density function (p.d.f.)*. The condition that the area under the curve is unity can be expressed mathematically as:

$$\int_{-\infty}^{\infty} f(E)\,\mathrm{d}E = 1$$

The probability that the error in any one particular measurement lies between two levels E_1 and E_2 can be calculated by measuring the area under the curve contained between two vertical lines drawn through E_1 and E_2, as shown by the right-hand hatched area in figure 7.2. This can be expressed mathematically as:

$$P(E_1 \leqslant E \leqslant E_2) = \int_{E_1}^{E_2} f(E)\,\mathrm{d}E \tag{7.4}$$

This expression (7.4) is often known as the *error function*.

Of particular importance for assessing the maximum error likely in any one measurement is the *cumulative distribution function (c.d.f.)*. This is defined as the probability of observing a value less than or equal to E_0, and is expressed mathematically as:

$$P(E \leqslant E_0) = \int_{-\infty}^{E_0} f(E)\, dE \qquad (7.5)$$

Thus the c.d.f. is the area under the curve to the left of a vertical line drawn through E_0, as shown by the left-hand hatched area on figure 7.2.

Three special types of frequency distribution known as the Gaussian, binomial and Poisson distributions exist, and these are very important because most data sets approach closely to one or other of them. The distribution of relevance to data sets containing random measurement errors is the Gaussian one.

Gaussian distribution

A Gaussian curve is defined as an error frequency distribution where the error frequency and magnitude are related by the expression:

$$f(E) = \frac{1}{\sigma(2\pi)^{\frac{1}{2}}} \exp\left[-(E-m)^2/2\sigma^2\right] \qquad (7.6)$$

where m is the mean value and the other quantities are as defined before. Most measurement data sets such as the mass values above in (7.2) fit to a Gaussian distribution curve because, if errors are truly random, small deviations from the mean value occur much more often than large deviations, i.e. the number of small errors is much larger than the number of large ones. Alternate names for the Gaussian distribution curve are the *normal distribution* or *bell-shaped distribution*.

The Gaussian distribution curve is symmetrical about the line of zero measurement error, which means that positive errors away from the mean value occur in equal quantities to negative errors in any data set containing measurements subject to random error. If the standard deviation is used as a unit of error, the curve can be used to determine what probability there is that the error in any particular measurement in a data set is greater than a certain value. By substituting the expression for $f(E)$ in (7.6) into the probability equation (7.4), it can be shown that, for Gaussian-distributed data values, 84.1% of the values lie within the bounds of $\pm\sigma$. Boundaries of $\pm2\sigma$ contain 97.7% of data points, and extending the boundaries to $\pm3\sigma$ encompasses 99.85% of data points. The probability of any data point lying outside particular error boundaries can therefore be expressed by the following table.

Error boundaries	*Data points within boundary*	*Probability of any particular data point being outside boundary*
$\pm \sigma$	84.1%	15.9%
$\pm2\sigma$	97.7%	2.3%
$\pm3\sigma$	99.85%	0.15%

Standard error of the mean

The foregoing analysis is only strictly true for measurement sets containing infinite populations. It is not of course possible to obtain an infinite number of data values, and some error must therefore be expected in the calculated mean value of the practical, finite data set available. If several subsets are taken from an infinite data population, then, by the *central limit theorem*, the means of the subsets will form a Gaussian distribution about the mean of the infinite data set. The error in the mean of a finite data set is usually expressed as the *standard error of the mean*, α, which is calculated as:

$$\alpha = \sigma/n^{1/2}$$

This tends toward zero as the number of measurements in the data set is expanded toward infinity. The value obtained from a set of n measurements $x_1, x_2, \ldots, x_n$ is then expressed as:

$$x = x_{mean} \pm \alpha$$

For the data set of mass measurements above in (7.2), $\sigma = 0.318$ and $\alpha = 0.073$. The mass can therefore be expressed as 81.18 ± 0.07.

7.2   Estimation of total measurement system errors

A measurement system often consists of several separate components, each of which is subject to systematic and/or random errors. Mechanisms have now been presented for quantifying the errors arising from each of these sources and therefore the total error at the output of each measurement system component can be calculated. What remains to be investigated is how the errors associated with each measurement system component combine together, so that a total error calculation can be made for the complete measurement system.

Error in a product

If the outputs y and z of two measurement system components are multiplied together, the product can be written as:

$$P = yz$$

If the possible error in y is $\pm ay$ and in z is $\pm bz$, then the maximum and minimum values possible in P can be written as:

$$P_{max} = (y + ay)(z + bz)$$
$$= yz + ayz + byz + aybz$$

$$P_{min} = (y - ay)(z - bz)$$
$$= yz - ayz - byz + aybz$$

For typical measurement system components with output errors of up to 1–2% in magnitude, $a<<1$ and $b<<1$ and terms in $aybz$ are negligible compared with the other terms. Therefore we have:

$$P_{max} = yz(1 + a + b) \qquad P_{min} = yz(1 - a - b)$$

Thus the error in the product P is $\pm(a + b)$.

Example
If the power in a circuit is calculated from measurements of voltage and current in which the calculated maximum errors are respectively $\pm1\%$ and $\pm2\%$, then the possible error in the calculated power value is $\pm3\%$.

Error in a quotient

If the output measurement y of one system component with possible error $\pm ay$ is divided by the output measurement z of another system component with possible error $\pm bz$, then the maximum and minimum possible values for the quotient can be written as:

$$Q_{max} = \frac{y + ay}{z - bz}$$
$$= \frac{(y + ay)(z + bz)}{(z - bz)(z + bz)}$$
$$= \frac{yz + ayz + byz + abyz}{z^2 - b^2 z^2}$$

$$Q_{min} = \frac{y - ay}{z + bz}$$
$$= \frac{(y - ay)(z - bz)}{(z + bz)(z - bz)}$$
$$= \frac{yz - ayz - byz + abyz}{z^2 - b^2 z^2}$$

For $a<<1$ and $b<<1$, terms in ab and b^2 are negligible compared with the other terms. Hence:

$$Q_{max} = \frac{yz(1 + a + b)}{z^2} \qquad Q_{min} = \frac{yz(1 - a - b)}{z^2}$$

i.e.

$$Q = \frac{y}{z} \pm \frac{y}{z}(a + b)$$

Thus the error in the quotient is $\pm(a + b)$.

Example
If the resistance in a circuit is calculated from measurements of voltage and current where the respective errors are $\pm1\%$ and $\pm0.3\%$, the likely error in the resistance value is $\pm1.3\%$.

Error in a sum

If the two outputs y and z of separate measurement system components are to be added together, we can write the sum as:

$$S = y + z$$

If the maximum errors in y and z are $\pm ay$ and $\pm bz$ respectively, we can express the maximum and minimum possible values of S as:

$$S_{max} = y + ay + z + bz \qquad S_{min} = y - ay + z - bz$$

or

$$S = y + z \pm(ay + bz)$$

This relationship for S is not convenient because in this form the error term cannot be expressed as a fraction or percentage of the calculated value for S, and it is also statistically incorrect because the errors in y and z are unlikely to both be at their maximum or minimum values at the same time.

Following statistical analysis, the most probable maximum error in S can be expressed by a quantity e, where e is given by:

$$e = [(ay)^2 + (bz)^2]^{1/2} \tag{7.7}$$

Thus:

$$S = (y + z) \pm e$$

This can be expressed in the alternative form:

$$S = (y + z)(1 \pm f) \tag{7.8}$$

where $f = e/(y + z)$.

Example
A circuit requirement for a resistance of 550 Ω is satisfied by connecting together two resistors of nominal values 220 Ω and 330 Ω in series. If each resistor has a tolerance of $\pm2\%$, the error in the sum calculated according to equations (7.7) and (7.8) is given by:

$$e = [(0.02 \times 220)^2 + (0.02 \times 330)^2]^{1/2} = 7.93$$

$$f = 7.93/550 = 0.0144$$

Thus the total resistance S can be expressed as:

$$S = 550 \ \Omega \pm 7.93 \ \Omega$$

or

$$S = 550(1 + 0.0144) \ \Omega$$

i.e.

$$S = 550 \ \Omega \pm 1.4\%.$$

Error in a difference

If the two outputs y and z of separate measurement systems are to be subtracted from one another, and the possible errors are $\pm ay$ and $\pm bz$, then the difference S can be expressed as:

$$S = (y - z) \pm e \quad \text{or} \quad S = (y - z)(1 \pm f)$$

where e is calculated as above (equation (7.7)) and $f = e/(y - z)$.

Example
A fluid flow rate is calculated from the difference in pressure measured on both sides of an orifice plate. If the pressure measurements are 10.0 bar and 9.5 bar and the error in the pressure-measuring instruments is specified as $\pm 0.1\%$, then values for e and f can be calculated as:

$$e = [(0.001 \times 10)^2 + (0.001 \times 9.5)^2]^{1/2} = 0.0138$$

$$f = 0.0138/0.5 = 0.0276$$

Thus the pressure difference can be expressed as 0.5 bar $\pm 2.8\%$. This example illustrates very poignantly the relatively large error which can arise when calculations are made based on the difference between two measurements.

7.3   Presentation of data

We have now discussed what the various sources of systematic and random errors are in a measurement system, how the magnitude of each error component is calculated, and how the overall measurement system accuracy is calculated from the individual error components. The output data being discussed might arise as part of a laboratory experiment or it might arise out of monitoring process variables in an industrial environment. In either case, there is a need to present the data in a useful format for future study and use. The alternate formats available are tabular and graphical ones, each of which has certain merits compared with the other. In many data collection exercises, part of the measurements and calculations are expressed in tabular form and part graphically, making best use of the merits of each technique.

Tabular data presentation

A tabular presentation allows data values to be recorded in a precise way which maintains exactly the accuracy to which the data values were measured. In other words, the data values are written down exactly as measured. Besides recording the raw data values as measured, tables often also contain further values calculated from the raw data.

An example of a tabular data presentation is given in table 7.1. This records the results of an experiment to determine the strain induced in a bar of material under a range of stresses. Data were obtained by applying a sequence of forces to the end of the bar and using an extensometer to measure the change in length. Values of the stress and strain in the bar are calculated from these measurements and also included in the table. The final row, which is of crucial importance in any tabular presentation, is the estimate of possible error in each calculated result.

A table of measurements and calculations should conform to several rules as illustrated in table 7.1.

Table 7.1 Sample tabular presentation of data: Table of measured applied forces and extensometer readings and calculations of stress and strain

Force applied (kN)	Extensometer reading (divisions)	Stress (N/m^2)	Strain
0	0	0	0
2	4.0	15.5	19.8×10^{-5}
4	5.8	31.0	28.6×10^{-5}
6	7.4	46.5	36.6×10^{-5}
8	9.0	62.0	44.4×10^{-5}
10	10.6	77.5	52.4×10^{-5}
12	12.2	93.0	60.2×10^{-5}
14	13.7	108.5	67.6×10^{-5}
Possible error in measurements ± 0.2	± 0.2	± 1.5	$\pm 1.0 \times 10^{-5}$

(i)   The table should have a title which explains what data are being presented within the table.

(ii)   Each column of figures in the table should refer to the measurements or calculations associated with one quantity only.

(iii)   Each column of figures should be headed by a title which identifies the data values contained in the column.

(iv)   The units in which quantities in each column are measured should be stated at the top of the column.

(v) Titles and columns should be separated by horizontal (and sometimes vertical) lines.

(vi) The errors associated with each data value quoted in the table should be given. The form shown in table 7.1 is a suitable way to do this when the error level is the same for all data values in a particular column. If error levels vary however, then it is preferable to write the error boundaries alongside each entry in the table.

Graphical presentation of data

Presentation of data in graphical form involves some compromise in the accuracy to which the data are recorded, as the exact values of measurements are lost. However, graphical presentation has two important advantages over tabular presentation.

(i) Graphs provide a pictorial representation of results which is more readily comprehended than a set of tabular results.

(ii) Graphs are particularly useful for expressing the quantitative significance of results and showing whether a linear relationship exists between two variables.

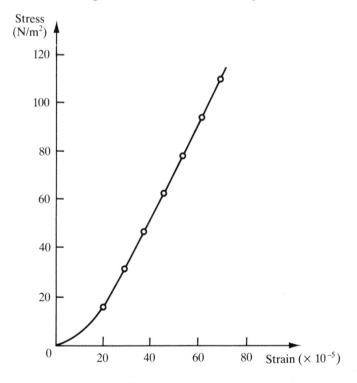

Figure 7.3   Sample graphical presentation of data: graph of stress against strain

Figure 7.3 shows a graph drawn from the stress and strain values given in table 7.1. Construction of the graph involves first of all marking the points corresponding to the stress and strain values. The next step is to draw some line through these data points which best represents the relationship between the two variables. This line will normally be either a straight one or a smooth curve. The data points will not usually lie exactly on this line but instead will lie on either side of it. The magnitude of the excursions of the data points from the line drawn will depend on the magnitude of the random measurement errors associated with the data.

As for tables, certain rules exist for the proper representation of data in graphical form:

(i)   The graph should have a title or caption which explains what data are being presented in the graph.
(ii) Both axes of the graph should be labeled to express clearly what variable is associated with each axis and to define the units in which the variables are expressed.
(iii) The number of points marked along each axis should be kept reasonably small – about five divisions is often a suitable number.
(iv) No attempt should be made to draw the graph outside the boundaries corresponding to the maximum and minimum data values measured, i.e. in figure 7.3, the graph stops at a point corresponding to the highest measured stress value of 108.5.

Fitting curves to data points on a graph

The procedure of drawing a straight line or smooth curve as appropriate which passes close to all data points on a graph, rather than joining the data points by a jagged line which passes through each data point, is justified on account of the random errors which are known to affect measurements. Any line between the data points is mathematically acceptable as a graphical representation of the data if the maximum deviation of any data point from the line is within the boundaries of the identified level of possible measurement errors. However, within the range of possible lines which could be drawn, only one will be optimum. This optimum line is where the sum of negative error data points on one side of the line is balanced by the sum of positive error data points on the other side of the line. The nature of the data points is often such that a perfectly acceptable approximation to the optimum can be obtained by drawing a line through the data points by eye. In other cases, however, it is necessary to fit a line by regression techniques.

Regression techniques

Regression techniques consist of finding a mathematical relationship between measurements of two variables y and x, such that the value of one variable y can be predicted from a measurement of the other variable x. This procedure is simplest if a straight-line relationship exists between the variables which can be estimated by linear least-squares regression.

In many cases, inspection of the raw data points plotted on a graph shows that a straight-line relationship between the points does not exist. However, knowledge of physical laws governing the data can often suggest a suitable alternate form of relationship between the two sets of variable measurements.

In some cases, the measured variables can be transformed such that a linear relationship is obtained. For example, suppose that two variables y and x are related according to:

$$y = ax^c$$

A linear relationship from this can be derived as:

$$\log(y) = \log(a) + c \log(x)$$

Thus if a graph is constructed of $\log(y)$ plotted against $\log(x)$, the parameters of a straight-line relationship can be estimated by linear least-squares regression.

Apart from linear relationships, physical laws often suggest a relationship where one variable y is related to another variable x by a power series of the form:

$$y = a_0 + a_1 x + a_2 x^2 + \ldots + a_p x^p$$

Estimation of the parameters a_0, ..., a_p is very difficult if p has a large value. Fortunately, a relationship where p only has a small value can be fitted to most data sets. Quadratic least-squares regression is used to estimate parameters where p has a value of 2, and for larger values of p polynomial least-squares regression is used for parameter estimation.

Where the appropriate form of relationship between variables in measurement data sets is not obvious either from visual inspection or from consideration of physical laws, a method which is effectively a trial-and-error one has to be applied. This consists of estimating the parameters of successively higher-order relationships between y and x until a curve is found which fits the data sufficiently closely. What level of closeness is acceptable is considered in the later section on confidence tests.

Linear least-squares regression

If a linear relationship between y and x exists for a set of n measurements $y_1, \ldots, y_n$, $x_1, \ldots, x_n$, then this relationship can be expressed as

$$y = a + bx$$

where the coefficients a and b are constants. The purpose of least-squares regression is to select the optimum values for a and b such the line gives a good fit to the measurement data.

The deviation of each point (x_i, y_i) from the line can be expressed as d_i, where

$$d_i = y_i - (a + bx_i)$$

The best-fit line is obtained when the sum of the squared deviations, S, is a minimum, i.e. when

$$S = \sum_{i=1}^{n} (d_i^2) = \sum_{i=1}^{n} (y_i - a - bx_i)^2$$

is a minimum. The minimum can be found by setting the partial derivatives $\partial S / \partial a$ and $\partial S / \partial b$ to zero and solving the resulting two simultaneous (normal) equations:

$$\partial S/\partial a = \Sigma \, 2(y_i - a - bx_i)(-1) = 0$$

$$\partial S/\partial b = \Sigma \, 2(y_i - a - bx_i)(-x_i) = 0$$

Using x_m and y_m to represent the mean values of x and y, the least-squares estimates $\hat{a}$ and $\hat{b}$ of the coefficients a and b can be written as:

$$\hat{b} = \frac{\Sigma (x_i - x_m)(y_i - y_m)}{\Sigma (x_i - x_m)^2} \tag{7.9}$$

$$\hat{a} = y_m - \hat{b} x_m \tag{7.10}$$

Example

In an experiment to determine the variation of the specific heat of a substance with temperature, the following table of results was obtained:

Temp. (°C)	40	45	50	55	60	65	70	75	80	85	90	95	100
Sp. heat	1.38	1.43	1.46	1.49	1.56	1.57	1.59	1.64	1.69	1.72	1.78	1.83	1.85

Fit a straight line to this data set using least-squares regression and estimate the specific heat at a temperature of 72°C.

Solution

Let y represent the specific heat and x represent the temperature. Then a suitable straight line is given by $y = a + bx$. We can now proceed to calculate estimates for the coefficients a and b using equations (7.9) and (7.10) above.

The first step is to calculate the mean values of x and y. These are found to be $x_m = 70$ and $y_m = 1.6146$.

Next, we need to tabulate $(x_i - x_m)$ and $(y_i - y_m)$ for each pair of data values:

i	$(x_i - x_m)$	$(y_i - y_m)$
1	−30	−0.2346
2	−25	−0.1846
.		
.		
.		
13	+30	+0.2354

Now, applying equations (7.9) and (7.10), we obtain:

$$\hat{b} = 0.007\,8177 \qquad \text{and} \qquad \hat{a} = 1.066\,923$$

i.e.

$$y = 1.066\,923 + 0.007\,8177x$$

Using this equation, at a temperature (x) of 72°C, the specific heat (y) is 1.63. Note that in this solution, we have only specified the answer to an accuracy of three figures, which is the same accuracy as the measurements. Any greater number of figures in the answer would be meaningless.

Quadratic least-squares regression

Quadratic least-squares regression is used to estimate the parameters of a relationship

$$y = a + bx + cx^2$$

between two sets of measurements $y_1, \ldots, y_n$ and $x_1, \ldots, x_n$.

The deviation of each point (x_i, y_i) from the line can be expressed as d_i, where

$$d_i = y_i - (a + bx_i + cx_i^2)$$

The best-fit line is obtained when the sum of the squared deviations, S, is a minimum, i.e. when

$$S = \sum_{i=1}^{n} (d_i^2) = \sum_{i=1}^{n} (y_i - a - bx_i - cx_i^2)^2$$

is a minimum. The minimum can be found by setting the partial derivatives $\partial S/\partial a$, $\partial S/\partial b$ and $\partial S/\partial c$ to zero and solving the resulting simultaneous equations, as for the linear least-squares regression case above. Standard computer programs to estimate the parameters a, b and c by numerical methods are widely available and therefore a detailed solution is not presented here.

Polynomial least-squares regression

Polynomial least-squares regression is used to estimate the parameters of the pth-order relationship

$$y = a_0 + a_1x + a_2x^2 + \ldots + a_px^p$$

between two sets of measurements $y_1, \ldots, y_n$ and $x_1, \ldots, x_n$.

The deviation of each point (x_i, y_i) from the line can be expressed as d_i, where

$$d_i = y_i - (a_0 + a_1x_1 + a_2x_i^2 + \ldots + a_px_i^p)$$

The best-fit line is obtained when the sum of the squared deviations given by

$$S = \sum_{i=1}^{n} (d_i^2)$$

is a minimum. The minimum can be found as before by setting the p partial derivatives $\partial S/\partial a_0$, ..., $\partial S/\partial a_p$ to zero and solving the resulting simultaneous equations. Again, as for the quadratic least-squares regression case, standard computer programs to estimate the parameters a_0, ..., a_p by numerical methods are widely available and therefore a detailed solution is not presented here.

Confidence tests in curve fitting

Having applied least-squares regression to estimate the parameters of a chosen relationship, some form of follow-up procedure is clearly required to assess how well the estimated relationship fits the data points. One fundamental requirement in curve fitting is that the maximum deviation d_i of any data point (y_i, x_i) from the fitted curve is less than the calculated maximum measurement error level. For some data sets, it is impossible to find a relationship between the data points which satisfies this requirement. This normally occurs when both variables in a measurement data set are subject to random variation, such as where the two sets of data values are measurements of human height and weight. Correlation analysis is applied in such cases to determine the degree of association between the variables.

Assuming that a curve can be fitted to the data points in a measurement set without violating the above fundamental requirement, a further simple curve-fitting confidence test is to calculate the sum of squared deviations S for the chosen y/x relationship and compare it with the value of S calculated for the next higher-order regression line which could be fitted to the data. Thus if a straight-line relationship is chosen, the value of S calculated should be of a similar magnitude to that obtained by fitting a quadratic relationship. If the value of S were substantially to a quadratic relationship, this would indicate that a quadratic relationship was a better fit to the data than a straight-line one and further tests would be needed to examine whether a cubic or higher-order relationship was a better fit still.

Other more sophisticated confidence tests exist such as the F-ratio test. However, these are outside the scope of this book.

Correlation tests

Where both variables in a measurement data set are subject to random fluctuations, correlation analysis is applied to determine the degree of association between the variables. For example, in the case already quoted of a data set containing measurements of human height and weight, we certainly expect some relationship between the variables of height and weight because a tall person is heavier *on average* than a short person. Correlation tests determine the strength of the relationship (or interdependence) between the measured variables, which is expressed in the form of a correlation coefficient.

For two sets of measurements $x_1, \ldots, x_n$ and $y_1, \ldots, y_n$ with means x_m and y_m, the correlation coefficient ϕ is given by:

$$\phi = \frac{\Sigma(x_i - x_m)(y_i - y_m)}{\{[\Sigma(x_i - x_m)^2][\Sigma(y_i - y_m)^2]\}^{1/2}}$$

The value of $|\phi|$ always lies between 0 and 1, with 0 representing the case where the variables are completely independent of one another and 1 the case where they are totally related to one another.

For $0 < |\phi| < 1$, linear least-squares regression can be applied to find relationships between the variables which allow x to be predicted from a measurement of y and y to be predicted from a measurement of x. This involves finding two separate regression lines of the form:

$$y = a + bx \quad \text{and} \quad x = c + dy$$

These two lines are not normally coincident as shown in figure 7.4.

Both lines pass through the centroid of the data points but their slopes are different. As $|\phi| \to 1$, the lines tend to coincidence, representing the case where the two variables are totally dependent upon one another. As $|\phi| \to 0$, the lines tend to orthogonal ones parallel to the x and y axes. In this case, the two sets of variables are totally independent. The best estimate of x given any measurement of y is $\bar{x}$ and the best estimate of y given any measurement of x is $\bar{y}$.

For the general case, the best fit to the data is that line which bisects the angle between the lines on figure 7.4.

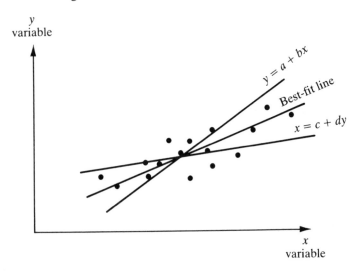

Figure 7.4 Relationship between two variables with random fluctuations

Exercises

7.1 In a survey of 15 owners of a certain model of car, the following figures for average petrol consumption were reported: 25.5, 30.3, 31.1, 29.6, 32.4, 39.4, 28.9, 30.0, 33.3, 31.4, 29.5, 30.5, 31.7, 33.0, 29.2. Calculate the mean value, the median value and the standard deviation of the data set.

7.2 What do you understand by the expressions: (a) probability density function; (b) central limit theorem.

Write down an expression for a Gaussian probability density function of given mean value π and standard deviation σ and show how you would obtain the best estimate of these two quantities from a sample of population n.

The following 10 measurements are made of the output voltage from a high-gain amplifier which is contaminated due to noise fluctuations: 1.53, 1.57, 1.54, 1.54, 1.50, 1.51, 1.55, 1.54, 1.56, 1.53. Determine the mean value and standard deviation. Hence estimate the accuracy to which the mean value is determined from these 10 measurements. If 1000 measurements were taken, instead of 10, but σ remained the same, by how much would the accuracy of the calculated mean value be improved?

7.3   A 3 V d.c. power source required for a circuit is obtained by connecting together two 1.5 V batteries in series. If the error in the voltage output of each battery is specified as $\pm 1\%$, calculate the possible error in the 3 V power source which they make up.

7.4   In order to calculate the heat loss through the wall of a building, it is necessary to know the temperature difference between the inside and outside walls. If temperature of 5°C and 20°C are measured on each side of the wall by mercury-in-glass thermometers with a range of -25°C to $+25$°C and a quoted accuracy figure of $\pm 1\%$ of full-scale reading, calculate the possible error in the calculated figure for the temperature difference.

7.5   The power dissipated in a car headlight is calculated by measuring the d.c. voltage drop across it and the current flowing through it ($P=VI$). If the possible errors in the measured voltage and current values are $\pm 1\%$ and $\pm 2\%$ respectively, calculate the possible error in the power value deduced.

7.6   The resistance of a carbon resistor is measured by applying a d.c. voltage across it and measuring the current flowing ($R = V/I$). If the voltage and current values are measured as 10 ± 0.1 V and 214 ± 5 mA respectively, calculate the value of the carbon resistor.

7.7   The quantities x and y are expected to be related in a linear fashion such that:

$$y = ax + b$$

Show that the best estimate of the constants a and b are given by:

$$a = \frac{\overline{x.y} - \bar{x}.\bar{y}}{\overline{x^2} - (\bar{x})^2}$$

$$b = \bar{y} - a.\bar{x}$$

Explain carefully the meaning of all the averages in the above two equations.

7.8   The characteristics of a chromel–constantan thermocouple is known to be approximately linear over the temperature range of 300–800°C. The output e.m.f. was measured practically at a range of temperatures and the following table of results obtained. Using least-squares regression, calculate the coefficients a and b for the relationship $T = aE + b$ which best describes the temperature – e.m.f. characteristic.

Temp. (°C)	300	325	350	375	400	425	450	475	500	525	550
e.m.f. (mV)	21.0	23.2	25.0	26.9	28.6	31.3	32.8	35.0	37.2	38.5	40.7

Temp. (°C)	575	600	625	650	675	700	725	750	775	800
e.m.f. (mV)	43.0	45.2	47.6	49.5	51.1	53.0	55.5	57.2	59.0	61.0

8

Measurement and Recording of Electrical Signals

As a large proportion of the measuring instruments in common use have an output in the form of an electrical signal, study of the devices available for measuring and recording such signals is very important. For some purposes, it is only necessary to have a visual indication of the signal level when monitoring the output of a measuring instrument, whilst for other purposes there is a requirement to record the output signal continuously for future study and use.

The instruments used to give a visual indication of the level of electrical signals are meters, of both analog and digital varieties, and the cathode ray oscilloscope. These will always be required for calibration procedures even if a signal is to be subsequently recorded continuously.

A wide range of instruments and techniques are available for recording electrical signals continuously (or pseudo-continuously). These include chart recorders, ultra-violet recorders, fiber optic recorders, magnetic tape recorders and data logging by a digital computer.

8.1 Meters

Meters exist in both digital and analog forms. The digital forms consist of various versions of the digital voltmeter (normally abbreviated to DVM), which are differentiated according to the method used to convert an analog voltage signal into a digital output reading. Analog meters can be divided into six subclassifications, according to their principles of operation. These six types are moving-coil, moving-iron, electrodynamic, induction, clamp-on and electrostatic.

Analog meters are electromechanical devices driving a pointer against a scale. They are prone to measurement errors from a number of sources, with quoted accuracy figures of between $\pm 0.1\%$ and $\pm 3\%$. Inaccurate scale marking during manufacture, bearing friction, bent pointers and variations in ambient temperature all limit measurement accuracy. Further human errors are introduced through parallax error (not reading the scale from directly above) and mistakes in interpolating between scale markings.

Digital meters give a reading in the form of a digital display. There are no problems of parallax and every observer sees the same value. In fact, digital meters are technically superior in every respect to analog meters and have quoted accuracy figures of between ±0.005% (measuring d.c. voltages) and ±2%. Unfortunately, this is only achieved at a considerably higher manufacturing cost than that involved in producing analog meters.

Additional advantages of digital voltmeters are their very high input impedance (10 MΩ compared with 1–20 kΩ for analog meters), the ability to measure signals of frequency up to 1 MHz and the common inclusion of features such as automatic ranging, which prevents overload and reverse polarity connection, etc.

Voltage-to-time conversion digital voltmeter

This is known as a ramp type of DVM. When an unknown voltage signal is applied to the input terminals of the instrument, a negative-slope ramp waveform is generated internally and compared with the input signal. When the two are equal, a pulse is generated which opens a gate, and at a later point in time a second pulse closes the gate when the negative ramp voltage reaches zero. The length of time between the gate opening and closing is monitored by an electronic counter, which produces a digital display according to the level of the input voltage signal. The best accuracy of this type is about ±0.05%.

Potentiometric digital voltmeter

This type of DVM uses the servo principle, where the error between the unknown input voltage level and a reference voltage is applied to a servo-driven potentiometer which adjusts the reference voltage until it balances the unknown voltage. The output reading is produced by a mechanical drum-type digital display driven by the potentiometer. This is a relatively cheap form of DVM.

Dual-slope integration digital voltmeter

In this type of DVM, the unknown voltage is applied to an integrator for a fixed time T_1, following which a reference voltage of opposite sign is applied to the integrator,

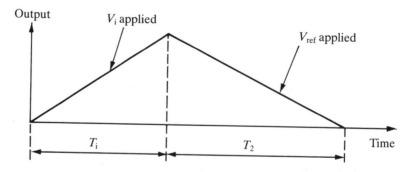

Figure 8.1   Output–time relationship for integrator in a dual-slope digital voltmeter (DVM)

which discharges down to a zero output in an interval T_2 measured by a counter. The output–time relationship for the integrator is shown in figure 8.1, from which the unknown voltage V_i can be calculated geometrically from the triangle as:

$$V_i = V_{ref}(T_1/T_2)$$

This is another relatively simple and cheap form of DVM.

Voltage-to-frequency conversion digital voltmeter

In this form, the unknown voltage signal is fed via a range switch and an amplifier to a converter circuit whose output is in the form of a train of voltage pulses at a frequency proportional to the magnitude of the input signal. The main advantage of this type of DVM is its ability to reject a.c. noise.

Moving-coil meters

A moving-coil meter is an analog instrument which responds only to direct current inputs and is shown schematically in figure 8.2. It consists of a rectangular coil wound round a soft iron core which is suspended in the field of a permanent magnet. The signal being measured is applied to the coil and this produces a radial magnetic field. Interaction between this induced field and the field produced by the permanent

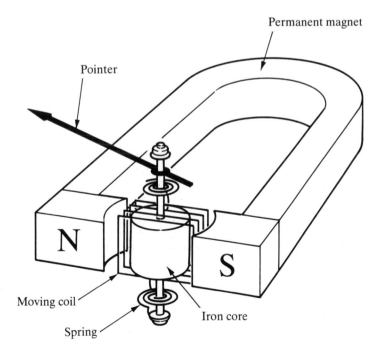

Figure 8.2   Moving-coil meter

magnet causes a torque, which results in rotation of the coil. The amount of rotation of the coil is measured by attaching a pointer to it which moves past a graduated scale.

The theoretical torque produced is given by:

$$T = BIhwN \qquad (8.1)$$

where B is the flux density of the radial field, I is the current flowing in the coil, h is the height of the coil, w is the width of the coil and N is the number of turns in the coil. If the iron core is cylindrical and the air gap between the coil and pole faces of the permanent magnet is uniform, then the flux density B is constant, and equation (8.1) can be rewritten as:

$$T = KI \qquad (8.2)$$

i.e. the torque is proportional to the coil current and the instrument scale is linear.

In certain situations, a non-linear scale such as a logarithmic one is required, which is achieved by using either a specially shaped core or specially shaped magnet pole faces.

Whilst figure 8.2 shows the traditional moving-coil instrument with a long U-shaped permanent magnet, many newer instruments employ much shorter magnets made from recently developed magnetic materials such as Alnico and Alcomax. These materials produce a substantially greater flux density, which, besides allowing the magnet to be smaller, has additional advantages in allowing reductions to be made in the size of the coil and in increasing the usable range of deflection of the coil to about 120°.

Moving-iron meters

This type of analog meter is suitable for measuring both d.c. and a.c. signals up to frequencies of 125 Hz. They are the cheapest form of meter available and consequently this type of meter is the one most commonly found in voltage and current measurement situations. The signal to be measured is applied to a stationary coil, and the associated field produced is often amplified by the presence of an iron structure associated with the fixed coil. The moving element in the instrument consists of an iron vane which is suspended within the field of the fixed coil. When the fixed coil is excited, the iron vane turns in a direction which increases the flux through it.

The majority of moving-iron instruments are either of the attraction type or of the repulsion type. A few instruments belong to a third combination type. The attraction type, where the iron vane is drawn into the field of the coil as the current is increased, is shown schematically in figure 8.3(a). The alternate repulsion type is sketched in figure 8.3(b).

For an excitation current I, the torque produced which causes the vane to turn is given by:

$$T = \frac{I^2}{2} \frac{dM}{d\theta}$$

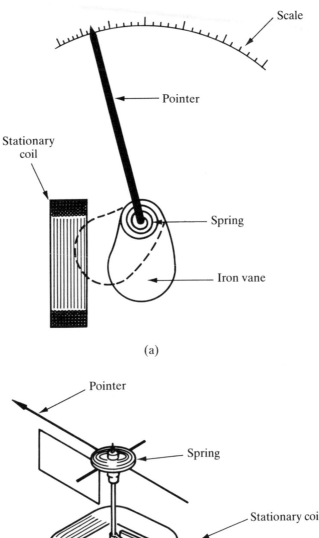

(a)

(b)

Figure 8.3 (a) Attraction-type moving-iron meter.
(b) Repulsion-type moving-iron meter

where M is the mutual inductance and θ is the angular deflection. Rotation is opposed by a spring which produces a backward torque given by:

$$T_s = K\theta$$

At equilibrium,

$$T = T_s$$

and θ is therefore given by:

$$\theta = \frac{I^2}{2K}\frac{dM}{d\theta}$$

The instrument thus has a square-law response where the deflection is proportional to the square of the signal being measured, i.e. the output reading is a root-mean-squared (r.m.s.) quantity.

Electrodynamic meters

Electrodynamic meters (or dynamometers) are analog instruments which can measure both d.c. quantities and a.c. quantities up to a frequency of 2 kHz. The typical construction of an electrodynamic meter is illustrated in figure 8.4. It consists of a circular moving coil mounted in the magnetic field produced by two separately wound circular stator coils which are connected in series. The torque is dependent

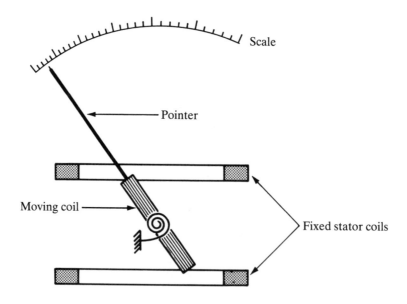

Figure 8.4   Electrodynamic meter

upon the mutual inductance between the coils and is given by:

$$T = I_1 I_2 \frac{dM}{d\theta} \tag{8.3}$$

where I_1 and I_2 are the currents flowing in the fixed and moving coils, M is the mutual inductance and θ represents the angular displacement between the coils.

Electrodynamic meters are typically expensive but have the advantage of being quite accurate. Voltage, current and power can all be measured if the fixed and moving coils are connected appropriately.

Induction meters

Induction meters are analog instruments which respond only to alternating current signals. The meter consists of two electromagnets arranged such that a phase difference θ exists between the fluxes ϕ_1 and ϕ_2 associated with the two coils, as sketched in figure 8.5. These two fluxes produce an induced e.m.f. and circulating concentric currents in the moving element of the instrument which consists of a conducting disk. The conducting disk tries to follow the rotating flux under the influence of a torque given by:

$$T = K\phi_1\phi_2 \sin \theta$$

For current and voltage measurement, the signal to be measured is applied to both electromagnet coils, which results in the meter having a square-law output

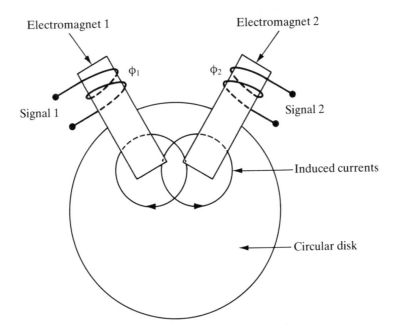

Figure 8.5   Induction meter

relationship of the form $T \propto I^2$ or $T \propto V^2$. The phase difference is obtained by shunting one coil with a non-inductive resistance, and rotation of the coil is opposed by a restraining spring.

When used as an induction-type energy meter, the disk in the instrument is unrestrained and free to rotate continuously. Circuit current is applied to one coil and circuit voltage to the other, thereby making the output reading a power measurement.

Clamp-on meters

These are used for measuring circuit currents and voltages in a non-invasive manner which avoids having to break the circuit being measured. The meter clamps onto a current-carrying conductor and the output reading is obtained by transformer action. The principle of operation is illustrated in figure 8.6, where it can be seen that the clamp-on jaws of the instrument act as a transformer core and the current-carrying conductor acts as a primary winding. Current induced in the secondary winding is rectified and applied to a moving-coil meter Although it is a very convenient instrument to use, the clamp-on meter has low sensitivity and the minimum current measurable is usually about 1 A.

Figure 8.6   Clamp-on meter

Electrostatic meters

Electrostatic meters use the principle of attraction between two oppositely charged circular plates, as illustrated in figure 8.7. The force produced is proportional to the square of the potential difference between the plates and is insufficient to provide an adequate deflection at voltages less than about 50 V. Electrostatic meters are there-

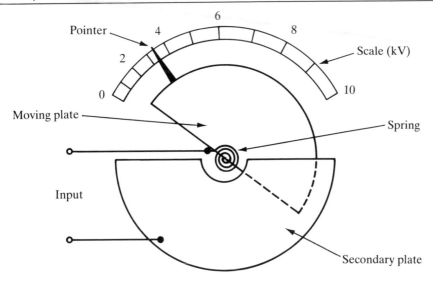

Figure 8.7   Electrostatic meter

fore normally only used for high-voltage applications, where voltages up to 500 kV can be measured directly. This type of instrument is very accurate, and it has the particular advantage of a very high input impedance, thereby drawing an extremely low current from the measured system.

8.2   Cathode ray oscilloscope

The cathode ray oscilloscope is probably the most versatile and useful instrument available for signal measurement. The more expensive models can measure signals at frequencies up to 500 MHz and even the cheapest models can measure signals up to 20 MHz. One particularly strong merit of the oscilloscope is its high input impedance, typically 1 MΩ, which means that the instrument has a negligible loading effect in most measurement situations. Some disadvantages of oscilloscopes include their fragility (being built around a cathode ray tube) and their moderately high cost. Also, whilst the accuracy of the best instruments is about $\pm 1\%$, the accuracy of the cheapest ones is only $\pm 10\%$. The most important aspects in the specification of an oscilloscope are its bandwidth, its risetime and its accuracy.

The bandwidth is defined as the range of frequencies over which the oscilloscope amplifier gain is within 3 dB[†] of its peak value, as illustrated in figure 8.8. The

---

[†] The decibel, commonly written dB, is used to express the ratio between two quantities. For two voltage levels V_1 and V_2, the difference between the two levels is expressed in decibels as:

$$20\log_{10}\left(\frac{V_1}{V_2}\right)$$

It following from this that:

$$20\log_{10}\left(\frac{0.707}{1}\right) = -3\,\text{dB}$$

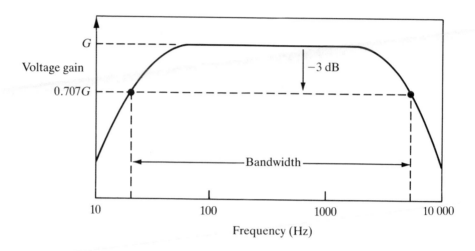

Figure 8.8   Bandwidth

−3 dB point is where the gain is 0.707 times its maximum value. In most oscillo-scopes, the amplifier is direct-coupled, which means that it amplifies d.c. voltages by the same factor as low-frequency a.c. ones. For such instruments, the minimum frequency measurable is zero and the bandwidth can be interpreted as the maximum frequency where the sensitivity (deflection per volt) is within 3 dB of the peak value. The oscilloscope must be chosen such that the maximum frequency to be measured is well within the bandwidth. The −3 dB specification means that an oscilloscope with a specified accuracy of ±2% and bandwidth of 100 MHz will only have an accuracy of ±5% when measuring 30 MHz signals and this accuracy will decrease still further at higher frequencies. Thus when applied to signal amplitude measurement, the oscilloscope is only usable at frequencies up to about 0.3 times its specified bandwidth.

The risetime is the transit time between the 10% and 90% levels of the response when a step input is applied to the oscilloscope. Oscilloscopes are normally designed such that:

$$\text{Bandwidth} \times \text{Risetime} = 0.35$$

Thus, for a bandwidth of 100 MHz, risetime = 0.35/100 000 000 = 3.5 ns.

An oscilloscope is a relatively complicated instrument which is constructed from a number of subsystems, and it is necessary to consider each of these in turn in order to understand how the complete instrument functions.

Cathode ray tube

The cathode ray tube is the fundamental part of an oscilloscope and is shown in figure 8.9. The cathode consists of a barium- and strontium-oxide-coated, thin, heated filament from which a stream of electrons are emitted. The stream of electrons is

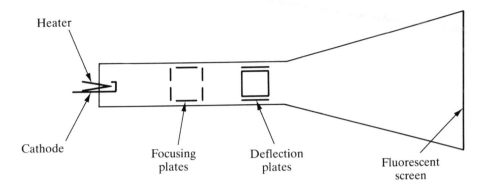

Figure 8.9   Cathode ray tube

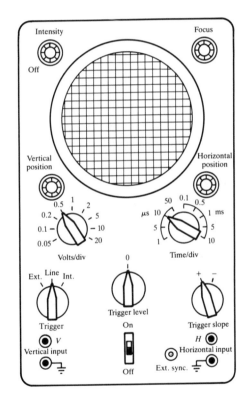

Figure 8.10   Controls of a simple oscilloscope

focused onto a well defined spot on a fluorescent screen by an electrostatic focusing system which consists of a series of metal disks and cylinders charged at various potentials. Adjustment of this focusing mechanism is provided by the 'Focusing/Astigmatism' controls on the front panel of an oscilloscope. An 'Intensity' control varies the cathode heater current and therefore the rate of emission of electrons, and thus adjusts the intensity of the display on the screen. These and other typical controls are shown in the illustration of the front panel of a simple oscilloscope given in figure 8.10.

Application of potentials to two sets of deflector plates mounted at right angles to one another within the tube provide for deflection of the stream of electrons, such that the spot where the electrons are focused on the screen is moved. The two sets of deflector plates are normally known as the horizontal and vertical deflection plates, according to the respective motion caused to the spot on the screen. The magnitude of any signal applied to the deflector plates can be calculated by measuring the deflection of the spot against a crossed-wires graticule etched on the screen.

In its most common mode of usage measuring time-varying signals, the unknown signal is applied, via an amplifier, to the y-axis (vertical) deflector plates and a time base to the x-axis (horizontal) deflector plates. In this mode of operation, the display on the oscilloscope screen is in the form of a graph with the magnitude of the unknown signal on the vertical axis and time on the horizontal axis.

Channels

The basic subsystem of an electron source, focusing system and deflector plates is often duplicated one or more times within the cathode ray tube to provide a capability of displaying two or more signals together on the screen. Each such basic subsystem is known as a channel, and therefore the commonest oscilloscope configuration having two channels can display two separate signals simultaneously.

Single-ended inputs

Most simple oscilloscopes have this type of input, which only allows signal voltages to be measured relative to ground. This type can be recognized by the presence of only one input terminal plus a ground terminal per oscilloscope channel.

Differential inputs

This type of input is provided on more expensive oscilloscopes and allows the potentials at two non-grounded points in a circuit to be compared. Two input terminals plus a ground terminal are provided for each channel. This arrangement can also be used in single-ended mode to measure a signal relative to ground by using just one of the input terminals plus ground.

Time base circuit

The purpose of a time base is to apply a voltage to the horizontal deflector plates such that the horizontal position of the spot is proportional to time. This voltage, in the form of a ramp known as a sweep waveform, must be applied repetitively, such that the motion of the spot across the screen appears as a straight line when a d.c. level is applied to the input channel. Furthermore, this time base voltage must be synchronized with the input signal in the general case of a time-varying signal, such that a steady picture is obtained on the oscilloscope screen. The length of time taken for the spot to traverse the screen is controlled by a 'Time/Div' switch, which sets the length of time taken by the spot to travel between two marked divisions on the screen, thereby allowing signals at a wide range of frequencies to be measured.

Each cycle of the sweep waveform is initiated by a pulse from a pulse generator. The input to the pulse generator is a sinusoidal signal known as a triggering signal, with a pulse being generated every time the triggering signal crosses a preselected slope and voltage level condition. This condition is defined by the 'Trigger level' and 'Trigger slope' switches. The former selects the voltage level on the trigger signal, commonly zero, at which a pulse is generated, whilst the latter selects whether pulsing occurs on a positive- or negative-going part of the triggering waveform.

Synchronization of the sweep waveform with the measured signal is most easily achieved by deriving the trigger signal from the measured signal, a procedure which is known as *internal triggering*. Alternatively, *external triggering* can be applied if the frequencies of the triggering signal and measured signals are related by an integer constant such that the display is stationary. External triggering is necessary when the amplitude of the measured signal is too small to drive the pulse generator, and it is also used in applications where there is a requirement to measure the phase difference between two sinusoidal signals of the same frequency. It is very convenient to use the 50 Hz line voltage for external triggering when measuring signals at mains frequency, and this is often given the name *line triggering*.

Vertical sensitivity control

In order to make the range of signal magnitudes measurable by an oscilloscope as wide as possible, a series of attenuators and pre-amplifiers are provided at the input. These condition the measured signal to the optimum magnitude for input to the main amplifier and vertical deflection plates. Selection of the appropriate input amplifier/ attenuator is made by setting a 'Volts/div' control associated with each oscilloscope channel. This defines the magnitude of the input signal which will cause a deflection of one division on the screen.

Display position control

The position on the screen at which a signal is displayed can be controlled in two ways. The horizontal position can be adjusted by a 'Horizontal position' knob on the oscilloscope front panel and similarly a 'Vertical position' knob controls the vertical position. These controls adjust the position of the display by biasing the measured signal with d.c. voltage levels.

8.3   Chart recorders

Chart recorders provide a relatively simple and cheap way of making permanent records of electrical signals. The two main kinds of chart recorder are the galvanometric type and the potentiometric type, with the former accounting for about 80% of the market.

Galvanometric recorders

These work on the same principle as a moving-coil meter except that the pointer draws an ink trace on paper, as illustrated in figure 8.11, instead of merely moving against a scale. The measured signal is applied to the coil, and the angular deflection

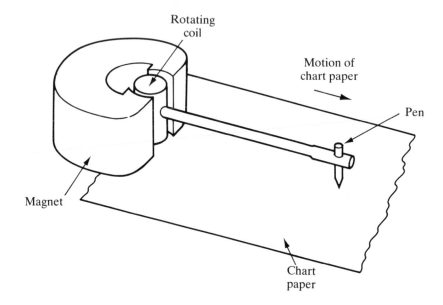

Figure 8.11   Simple galvanometric recorder

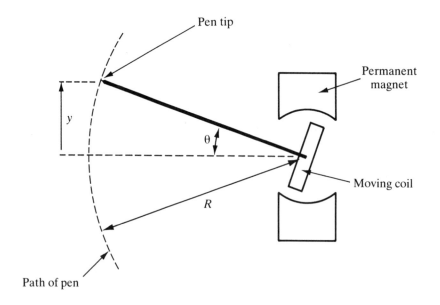

Figure 8.12 y vs θ relationship for standard chart recorder

of this and its attached pointer is proportional to the magnitude of the signal applied. The pointer normally carries a pen, and, by using a motor running at constant speed to drive the chart paper which the pointer is in contact with, a time history of the measured signal is obtained.

Inspection of figure 8.12 shows that the displacement y of the pen across the chart recorder is given by $y = R \sin \theta$. This sine relationship between the input signal and the displacement y is non-linear, and results in an error of 0.7% for deflections of $\pm 10°$. A more serious problem resulting from the pen moving in an arc is that it is difficult to relate the magnitude of deflection with the time axis. One way of overcoming this is to print a grid on the chart paper in the form of circular arcs, as illustrated in figure 8.13. Unfortunately, measurement errors often occur in reading this type of chart, as interpolation for points drawn between the curved grid lines is

Direction of
paper travel

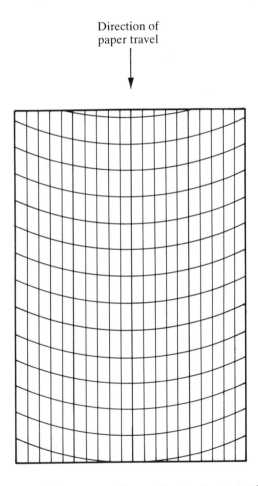

Figure 8.13   Curvilinear paper for standard galvanometric recorder

difficult. An alternate solution is to use heat-sensitive chart paper directed over a knife edge, and to replace the pen by a heated stylus, as illustrated in figure 8.14. The

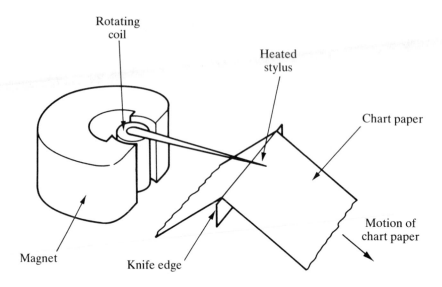

Figure 8.14   Knife-edge galvanometric recorder

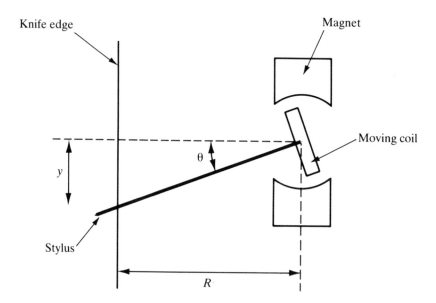

Figure 8.15   *y* vs θ relationship for knife-edge recorder

input–output relationship is still non-linear, with the deflection *y* being proportional
to tan θ as shown in figure 8.15, and the reading error for excursions of ±10° is still
0.7%. However, the rectilinearly scaled chart paper now required, as shown in figure
8.16, allows much easier interpolation between grid lines.

   Galvanometric recorders have a typical accuracy of ±2% and a resolution of
±1%. Measurement with galvanometric chart recorders is limited to signals below

Figure 8.16   Rectilinear chart paper for knife-edge galvanometric recorder

about 30 Hz in frequency, the reason for which is readily understood if their dynamic behavior is examined.

Dynamic behavior of galvanometric recorders

Neglecting friction, the torque equation with the recorder in steady state can be expressed as:

Torque due to current in coil = Torque due to spring

Following a step input, we can write:

Torque due to current in coil = Torque due to spring + Accelerating torque

or:

$$K_i I = K_s \theta + J\ddot{\theta} \tag{8.4}$$

where I is the coil current, θ is the angular displacement, J is the moment of inertia and K_i and K_s are constants.

Consider now what happens if a recorder with resistance R_r is connected to a transducer with resistance R_t and output voltage V_t, as shown in figure 8.17. The current flowing in steady state is given by:

$$I = V_t/(R_t+R_r)$$

Figure 8.17   Connection of transducer to chart recorder

When the transducer voltage V_t is first applied to the recorder coil, the coil will accelerate and, because the coil is moving in a magnetic field, a backward voltage will be induced in it given by:

$$V_i = -K_i\dot{\theta}$$

Hence the coil current is now given by:

$$I = \frac{V_t - K_i\dot{\theta}}{R_t + R_r}$$

Now substituting for I in the system equation (8.4):

$$K_i\left(\frac{V_t - K_i\dot{\theta}}{R_t + R_r}\right) = K_s\theta + J\ddot{\theta}$$

or rearranging:

$$\ddot{\theta} + \frac{K_i^2\dot{\theta}}{J(R_t+R_r)} + \frac{K_s\theta}{J} = \frac{K_iV_t}{J(R_t+R_r)} \qquad (8.5)$$

This is the standard equation of a second-order dynamic system, with natural frequency ω_n and damping factor β given by:

$$\omega_n = \frac{K_s}{J} \qquad \beta = \frac{K_i^2}{2K_sJ(R_t+R_r)}$$

In steady state, $\ddot{\theta} = \dot{\theta} = 0$ and equation (8.5) reduces to:

$$\frac{\theta}{V_t} = \frac{K_i}{K_s(R_t+R_r)} \qquad (8.6)$$

which is an expression describing the measurement sensitivity of the system.

The typical frequency response of a chart recorder for different damping factors is shown in figure 8.18. This shows that the best bandwidth is obtained at a

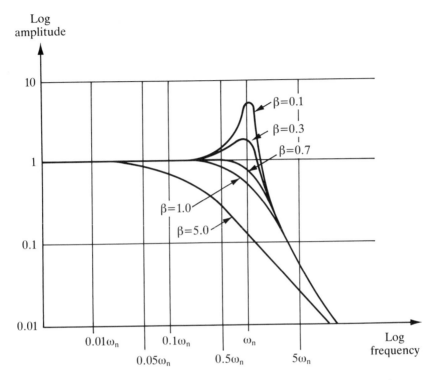

Figure 8.18   Frequency response of galvanometric chart recorder

damping factor of 0.7, and this is therefore the target when the instrument is designed. The damping factor depends not only on the coil and spring constants (K_i and K_s) but also on the total circuit resistance ($R_t + R_r$). Adding a series or parallel resistance between the transducer and recorder as illustrated in figure 8.19 respectively reduces or increases the damping factor. Consideration of the sensitivity expression of (8.6) shows, however, that any reduction in the damping factor takes place at the expense of a reduction in measurement sensitivity. Other methods to alter the damping factor are therefore usually necessary and these techniques include decreasing the spring constant and system moment of inertia. There is a limit as to how far the dynamic performance of a chart recorder can be improved, however, and other instruments such as ultra-violet recorders have to be used for higher-frequency signals.

Potentiometric recorders

Potentiometric recorders have much better specifications than galvanometric recorders, with an accuracy of ±0.1% of full scale and measurement resolution of ±0.2% f.s. being achievable. Such instruments employ a servo system, as shown in

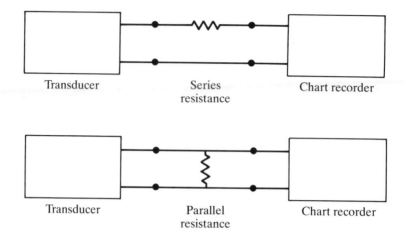

Figure 8.19   Addition of series and parallel resistances between transducer and chart recorder

figure 8.20, in which the pen is driven by a servomotor, and a potentiometer on the pen feeds back a signal proportional to pen position. This position signal is compared with the measured signal, and the difference is applied as an error signal which drives the motor. A consequence of this electromechanical balancing mechanism, however, is to give the instrument a slow response time in the range of 0.2–2.0 s. This means that potentiometric recorders are only suitable for measuring d.c. and slowly time-varying signals.

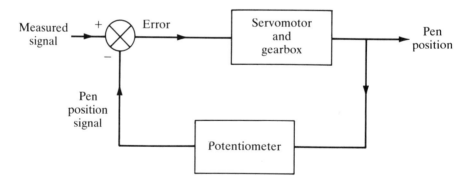

Figure 8.20   Servo system of potentiometer recorder

8.4   Ultra-violet recorders

The earlier discussion about galvanometric recorders concluded that restrictions on how far the system moment of inertia and spring constants can be reduced limited the maximum bandwidth to about 100 Hz. Ultra-violet recorders work on very similar principles to standard galvanometric chart recorders, but achieve a very significant reduction in system inertia and spring constants by mounting a narrow mirror rather

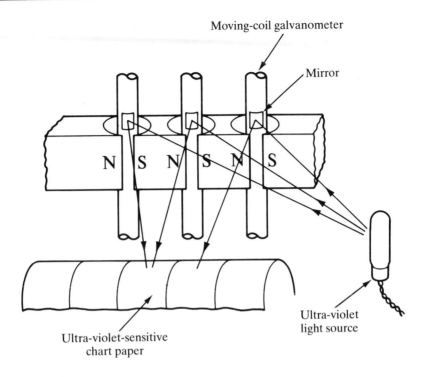

Figure 8.21 Ultra-violet recorder

than a pen system on the moving coil. This mirror reflects a beam of ultra-violet light onto ultra-violet-sensitive paper. It is usual to find several of these mirror–galvanometer systems mounted in parallel within one instrument to provide a multi-channel recording capability, as illustrated in figure 8.21. This arrangement enables signals at frequencies up to 13 kHz to be recorded with a typical accuracy of ±2% f.s. Whilst it is possible to obtain satisfactory permanent signal recordings by this method, special precautions are necessary to protect the u.v.-sensitive paper from light before use and to spray a fixing lacquer on it after recording. Such instruments must also be handled with extreme care, because the mirror galvanometers and their delicate mounting systems are easily damaged by relatively small shocks.

8.5 Fiber optic recorders

The fiber optic recorder is a very recent development which uses a fiber optic system to direct light onto light-sensitive paper. Fiber optic recorders are similar to oscilloscopes in construction, insofar as they have an electron gun and focusing system which directs a stream of electrons onto one point on a fluorescent screen. The screen is usually a long thin one instead of the square type found in an oscilloscope and only one set of deflection plates are provided. The signal to be recorded is applied to the deflection plates and the movement of the focused spot of electrons on the screen is proportional to the signal amplitude. A narrow strip of fiber optics in contact with the fluorescent screen transmits the motion of the spot to photosensitive paper held in close proximity to the other end of the fiber optic strip. By driving the photosensitive paper at a constant speed past the fiber optic strip, a time history of the measured signal is obtained. Such recorders are much more expensive than ultra-violet recorders but have an even higher bandwidth.

Whilst the construction above is the more common in fiber optic recorders, a second type also exists which uses a conventional square screen instead of a long thin one. This has a square faceplate attached to the screen housing a square array of fiber optics. The other side of the fiber optic system is in contact with chart paper. The effect of this is to provide a hard copy of the typical form of display obtainable on a cathode ray oscilloscope.

8.6   Magnetic tape recorders

Magnetic tape recorders can record analog signals up to 80 kHz in frequency. As the speed of the tape transport can be switched between several values, signals can be recorded at high speed and replayed at a lower speed. Such time scaling of the recorded information allows a hard copy of the signal behavior to be obtained on instruments such as ultra-violet and galvanometric recorders whose bandwidth is insufficient to allow direct signal recording. A 200 Hz signal cannot be recorded directly on a chart recorder, but if it is recorded on a magnetic tape recorder running at high speed and then replayed at a speed 10 times lower, its frequency will be time-scaled to 20 Hz which can be recorded on a chart recorder. Instrumentation tape recorders typically have between four and 10 channels, allowing many signals to be recorded simultaneously.

The two basic types of analog tape recording technique are direct recording and frequency-modulated recording. Direct recording offers the best data bandwidth but the accuracy of signal amplitude recording is quite poor, and this seriously limits the usefulness of this technique in most applications. The frequency-modulated technique offers much better amplitude recording accuracy, with an accuracy of ±5% at signal frequencies of 80 kHz. In consequence, this technique is very much more common than direct recording.

8.7   Computer data logging

The only technique available for recording signals at frequencies higher than 80 kHz is to use a digital computer. As the signals to be recorded are in analog form, a pre-requisite for this is an analog-to-digital converter board within the computer to sample the analog signals and convert them to the digital form in which the computer operates. The relevant aspects of the computer hardware involved have already been covered in chapter 3. Careful choice of the sampling interval is necessary so that an accurate digital record of the signal is obtained as explained in chapter 5, so that problems of aliasing, etc., are not encountered. Some prior analog signal conditioning may be required in some circumstances when a computer is used for data logging, again as mentioned in chapter 5.

Exercises

8.1 Discuss the dual-ramp technique of analog-to-digital conversion used in some types of digital voltmeter.

8.2 The waveform of a cyclic current is shown in figure 8.22. The current is measured using (a) a moving-coil ammeter and (b) a moving-iron ammeter. Calculate the readings on the two instruments.

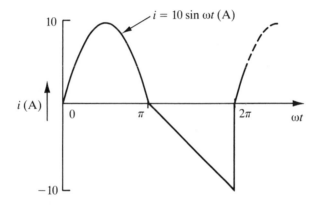

Figure 8.22   Waveform for exercise 8.2

8.3 (a) A moving-iron ammeter and a simple moving-coil ammeter are placed in series with a load in a half-wave rectifier circuit. The reading on the moving-iron instrument is 5 A. What is the reading on the other instrument?
(b) Calculate the readings on the two instruments if the other half-wave is also rectified. (Assume sinusoidal waveforms throughout.)

8.4 Describe the construction and operating principles of a repulsion-type moving-iron meter.
A compound voltage consisting of a d.c. level component of 5 V, together with a superimposed half-wave rectified sinusoidal waveform of peak value 4 V, occurs at a certain point in an electrical circuit. What would be the reading on (a) a moving-coil ammeter and (b) a moving-iron ammeter when used to measure this voltage waveform?

8.5 If the waveform shown in figure 8.23 was fed to (a) a true r.m.s. meter, (b) an average-measuring/r.m.s.-indicating meter, and (c) a peak-measuring/r.m.s.-indicating meter, what would the reading be on each instrument?

8.6 Describe the construction and operating principles of a dynamometer ammeter.
A dynamometer ammeter, a d.c. moving-coil ammeter and a 200 Ω resistor are connected in series with a rectifying device across a 240 V r.m.s. alternating sinusoidal supply. The rectifier can be considered to behave as a resistance of 100 Ω to current in one direction and of 1 kΩ to that in the opposite direction. Calculate (a) the readings on the two ammeters and (b) the power dissipated in the 200 Ω resistor.

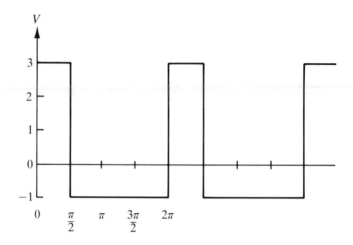

Figure 8.23   Waveform for exercise 8.5

8.7   The input impedance of an oscilloscope is 1 MΩ in parallel with 30 pF. Find the frequency at which this impedance falls to a magnitude one-tenth that at d.c.

8.8   (a) Explain the derivation of the expression

$$\ddot{\theta} + \frac{K_i^2\dot{\theta}}{JR} + \frac{K_s\theta}{J} = \frac{K_iV_t}{JR}$$

describing the dynamic response of a chart recorder following a step change in the electrical voltage output of a transducer connected to its input. Explain also what all the terms in the expression stand for. (Assume that the impedances of both the transducer and recorder have a resistive component only and that there is negligible friction in the system.)
(b) Derive expressions for the measuring system natural frequency, ω_n, the damping factor, β, and the steady-state sensitivity.
(c) Explain simple ways of increasing and decreasing the damping factor and describe the corresponding effect on measurement sensitivity.
(d) What damping factor gives the best system bandwidth?
(e) What aspects of the design of a chart recorder would you modify in order to improve the system bandwidth? What is the maximum bandwidth typically attainable in chart recorders, and if such a maximum-bandwidth instrument is available, what is the highest-frequency signal that such an instrument would be generally regarded as being suitable for measuring if the accuracy of the signal amplitude measurement is important?

INSTRUMENTS FOR MEASUREMENT OF PARTICULAR PHYSICAL VARIABLES

This second part of the book discusses the range of instruments which are available for measuring various physical quantities. In presenting each range of instruments, the aim has been to be as comprehensive as possible, to make the book useful as a future source of reference when choosing an instrument for any particular measurement situation. This method of treatment means that quite rare and expensive instruments as well as very common and cheap instruments are included, and therefore it is necessary to choose from the instrument ranges presented with care.

The first step in choosing an instrument for a particular measurement situation is to specify the static and dynamic characteristics required. These must always be defined carefully, as a high degree of accuracy, sensitivity, etc., in the instrument specification inevitably involves a high cost. The characteristics specified should therefore be the minimum necessary to achieve the required level of performance in the system that the instrument is providing a measurement for. Once the static and dynamic characteristics required have been defined, the range of instruments presented in each of the following chapters can be reduced to a subset containing those instruments which satisfy the defined specifications.

The second step in choosing an instrument is to consider the working conditions in which the instrument will operate. Conditions demanding special consideration are those where the instrument will be exposed to mechanical shocks, vibration, fumes, dust or fluids. Such considerations subdivide further the subset of possible instruments already identified. If at this stage a choice still exists, then the final criterion is one of cost.

The relevant criteria in instrument choice for measuring particular physical quantities are considered further in the following chapters.

9

Translational Displacement
Measurement

Translational displacement transducers are instruments which measure the motion of a body in a straight line between two points. Apart from their use as a primary transducer measuring the motion of a body, translational displacement transducers are also widely used as a secondary component in measurement systems, where some other physical quantity such as pressure, force, acceleration or temperature is translated into a translational motion by the primary measurement transducer.

Many different types of translational displacement transducer exist and these, along with their relative merits and characteristics, are discussed in the following sections of this chapter. The factors governing the choice of a suitable type of instrument in any particular measurement situation are considered in the final section at the end of the chapter.

9.1   Resistive potentiometer

The resistive potentiometer consists of a resistance element and a movable contact as shown in figure 9.1. A voltage V_s is applied across the two ends A and B of the resistance element and an output voltage V_0 is measured between the point of contact C of the sliding element and the end of the resistance element A. A linear relationship exists between the output voltage V_0 and the distance AC, which can be expressed by:

$$\frac{V_0}{V_s} = \frac{AC}{AB} \qquad (9.1)$$

The body whose motion is being measured is connected to the sliding element of the potentiometer, so that translational motion of the body causes a motion of equal magnitude of the slider along the resistance element and a corresponding change in the output voltage V_0.

Three different types of potentiometer exist, wire-wound, carbon-film and plastic-film, so named according to the material used to construct the resistance element. Wire-wound potentiometers consist of a coil of resistance wire wound on a

137

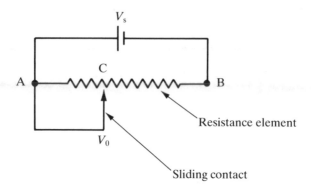

Figure 9.1   The resistive potentiometer

non-conducting former. As the slider moves along the potentiometer track, it makes contact with successive turns of the wire coil. This limits the resolution of the instrument to the distance from one coil to the next. Much better measurement resolution is obtained from potentiometers using either a carbon film or a conducting plastic film for the resistance element. The resolution of these is theoretically limited only by the grain size of the particles in the film, suggesting that measurement resolutions up to 10^{-4} ought to be attainable. In practice, the resolution is limited by mechanical difficulties in constructing the spring system which maintains the slider in contact with the resistance track, although these types are still considerably better than wire-wound types.

Operational problems of potentiometers all occur at the point of contact between the sliding element and the resistance track. The most common problem is dirt under the slider, which increases the resistance and thereby gives a false output voltage reading, or in the worst case causes a total loss of output. High-speed motion of the slider can also cause the contact to bounce, giving an intermittent output. Friction between the slider and the track can also be a problem in some measurement systems where the body whose motion is being measured is moved by only a small force of a similar magnitude to these friction forces.

The life expectancy of potentiometers is normally quoted as a number of reversals, i.e. as the number of times the slider can be moved backward and forward along the track. The figures quoted for wire-wound, carbon-film and plastic-film types are respectively 1 million, 5 million and 30 million. In terms of both life expectancy and measurement resolution, therefore, the carbon- and plastic-film types are clearly superior, although wire-wound types do have one advantage in respect of their lower temperature coefficient. This means that wire-wound types exhibit much less variation in their characteristics in the presence of varying ambient temperature conditions.

A typical accuracy figure which is quoted for translational-motion resistive potentiometers is $\pm 1\%$ of full-scale reading. Manufacturers produce potentiometers to cover a large span of measurement ranges. At the bottom end of this span, instruments with a range of ± 2 mm are available, whilst at the top end, instruments with a range of ± 1 m are produced.

The resistance of the instrument measuring the output voltage at the potentiometer slider can effect the value of the output reading, as discussed in chapter 6. As the slider moves along the potentiometer track, the ratio of the measured resistance to that of the measuring instrument varies, and thus the linear relationship between the measured displacement and the voltage output is distorted as well. This effect is minimized when the potentiometer resistance is small relative to that of the measuring instrument. This is achieved first by using a very high-impedance measuring instrument and secondly by keeping the potentiometer resistance as small as possible. Unfortunately, the latter is incompatible with achieving a high measurement sensitivity since this requires a high potentiometer resistance. A compromise between these two factors is therefore necessary. The alternate strategy of obtaining high measurement sensitivity by keeping the potentiometer resistance low and increasing the excitation voltage is not possible in practice because of the power rating limitation. This restricts the allowable power loss in the potentiometer to its heat dissipation capacity.

The process of choosing the best potentiometer from a range of instruments which are available, taking into account power rating and measurement linearity considerations, is illustrated in the example below.

Example

The output voltage from a translational-motion potentiometer of stroke length 0.1 m is to be measured by an instrument whose resistance is 10 kΩ. The maximum measurement error, which occurs when the slider is positioned two-thirds of the way along the element (i.e. when AC = 2AB/3 in figure 9.1), must not exceed 1% of the full-scale reading. The highest possible measurement sensitivity is also required. A family of potentiometers having a power rating of 1 W per 0.01 m and resistances ranging from 100 Ω to 10 kΩ in 100 Ω steps are available. Choose the most suitable potentiometer from this range and calculate the sensitivity of measurement which it gives.

Solution

Referring to the lettering used in figure 9.1, let the resistance of portion AC of the resistance element be R_i and that of the whole length AB of the element be R_t. Also, let the resistance of the measuring instrument be R_m and the output voltage measured by it be V_m. When the voltage-measuring instrument is connected to the potentiometer, the net resistance across AC is the sum of two resistances in parallel (R_i and R_m) given by:

$$R_{AC} = \frac{R_i R_m}{R_i + R_m}$$

Let the excitation voltage applied across the ends AB of the potentiometer be V and the resultant current flowing between A and B be I. Then I and V are related by:

$$I = \frac{V}{R_{AC} + R_{CB}} = \frac{V}{[R_i R_m/(R_i + R_m)] + R_t - R_i}$$

V_m can now be calculated as:

$$V_m = IR_{AC} = \frac{VR_iR_m}{\{[R_iR_m/(R_i+R_m)]+R_t-R_i\}(R_i+R_m)}$$

If we express the voltage which exists across AC in the absence of the measuring instrument as V_0, then we can express the error due to the loading effect of the measuring instrument as:

$$\text{Error} = V_0 - V_m$$

From equation (9.1)

$$V_0 = R_iV/R_t$$

Thus,

$$\text{Error} = V_0 - V_m = V\left(\frac{R_i}{R_t} - \frac{R_iR_m}{\{[R_iR_m/(R_i+R_m)]+R_t-R_i\}(R_i+R_m)}\right)$$

$$= V\left(\frac{R_i^2(R_t-R_i)}{R_t(R_iR_t+R_mR_t-R_i^2)}\right) \tag{9.2}$$

Substituting $R_i=2R_t/3$ into equation (9.2) to find the maximum error:

$$\text{Maximum error} = \frac{2R_t}{2R_t + 9R_m}$$

For a maximum error of 1%:

$$\frac{2R_t}{2R_t + 9R_m} = 0.01 \tag{9.3}$$

Substituting $R_m = 10\,000$ into the above expression (9.3) gives $R_t = 454\ \Omega$. The nearest resistance values in the range of potentiometers available are $400\ \Omega$ and $500\ \Omega$. The value of $400\ \Omega$ has to be selected as this is the only one which gives a maximum measurement error of less than 1%.

The thermal rating of the potentiometers is quoted as 1 W/0.01 m, i.e. 10 W for the total length of 0.1 m. By Ohm's law,

$$\text{Maximum supply voltage} = (\text{Power} \times \text{Resistance})^{1/2} = (10\times400)^{1/2} = 63.25\ \text{V}$$

Thus

$$\text{Measurement sensitivity} = 63.25/0.1\ \text{V/m} = 632.5\ \text{V/m}$$

9.2   Linear variable differential transformer (LVDT)

The linear variable differential transformer, which is commonly known by the abbreviation LVDT, consists of a transformer with a single primary winding and two secondary windings connected in the series opposing manner shown in figure 9.2. The object whose translational displacement is to be measured is physically attached to the central iron core of the transformer, so that all motions of the body are transferred to the core.

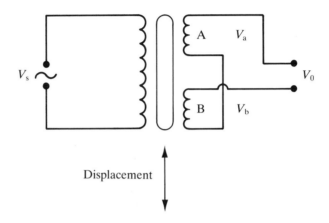

Figure 9.2   The linear variable differential transformer (LVDT)

For an excitation voltage V_s given by

$$V_s = V_p \sin(\omega t)$$

the e.m.f.'s induced in the secondary windings V_a and V_b are given by:

$$V_a = K_a \sin(\omega t - \phi) \qquad V_b = K_b \sin(\omega t - \phi)$$

The parameters K_a and K_b depend on the amount of coupling between the respective secondary and primary windings and hence on the position of the iron core. With the core in the central position, $K_a = K_b$, and we have

$$V_a = V_b = K \sin(\omega t - \phi)$$

Because of the series opposition mode of connection of the secondary windings, $V_0 = V_a - V_b$, and hence with the core in the central position, $V_0 = 0$.

Suppose now that the core is displaced upward (i.e. toward winding A) by a distance x. If then $K_a = K_1$ and $K_b = K_2$, we have:

$$V_0 = (K_1 - K_2) \sin(\omega t - \phi)$$

If, alternately, the core were displaced downward from the null position (i.e. toward winding B) by a distance x, the values of K_a and K_b would then be $K_a = K_2$ and $K_b = K_1$, and we would have:

$$V_0 = (K_2 - K_1)\sin(\omega t - \phi)$$
$$= (K_1 - K_2)\sin[\omega t + (\pi - \phi)]$$

Thus for equal-magnitude displacements $+x$ and $-x$ of the core away from the central (null) position, the magnitude of the output voltage V_0 is the same in both cases. The only information about the direction of movement of the core is contained in the phase of the output voltage, which differs between the two cases by 180°. If, therefore, measurements of core position on both sides of the null position are required, it is necessary to measure the phase as well as the magnitude of the output voltage. The relationship between the magnitude of the output voltage and the core position is approximately linear over a reasonable range of movement of the core on either side of the null position and is expressed using a constant of proportionality C as:

$$V_0 = Cx$$

The only moving part in an LVDT is the central iron core. As the core is only moving in the air gap between the windings, there is no friction or wear during operation. For this reason, the instrument is a very popular one for measuring linear displacements and has a quoted life expectancy of 200 years. The typical accuracy is $\pm 0.5\%$ of full-scale reading and measurement resolution is almost infinite. Instruments are available to measure a wide span of measurements from ± 100 μm to ± 100 mm. The instrument can be made suitable for operation in corrosive environments by enclosing the windings within a non-metallic barrier, which leaves the magnetic flux paths between the core and windings undisturbed. An epoxy resin is commonly used to encapsulate the coils for this purpose. One further operational advantage of the instrument is its insensitivity to mechanical shock and vibration.

Some problems which affect the accuracy of the LVDT are the presence of harmonics in the excitation voltage and stray capacitances, both of which cause a non-zero output of low magnitude when the core is in the null position. It is also impossible in practice to produce two identical secondary windings, and the small asymmetry which invariably exists between the secondary windings adds to this non-zero null output. The magnitude of this is always less than 1% of the full-scale output and in many measurement situations is of little consequence. If necessary, this non-zero null output can be eliminated by several means (see Doebelin 1975).

9.3 Variable-inductance transducers

Variable-inductance transducers, which are also sometimes referred to as variable-reluctance or variable-permeance transducers, exist in two separate forms. Both of these rely for their operation on the variation of the mutual inductance between two magnetically coupled parts. In this respect they are similar to the LVDT described in

(a)

(b)

Figure 9.3   Variable-inductance transducers

the last section. However, whereas the LVDT has three windings, both types of variable-inductance transducer have only one winding.

The two types of variable-inductance transducer are shown in figure 9.3. In the first of these (figure 9.3(a)), the single winding on the central limb of an 'E'-shaped ferromagnetic body is excited with an alternating voltage. The displacement to be measured is applied to a ferromagnetic plate in close proximity to the 'E' piece. Movements of the plate alter the flux paths and hence cause a change in the current flowing in the winding. By Ohm's law, the current flowing in the winding is given by:

$$I = V/\omega L$$

For fixed values of ω and V, this equation becomes:

$$I = 1/KL$$

where K is a constant. The relationship between L and the displacement (d) applied to the plate is a non-linear one, and hence the output current–displacement characteristic has to be calibrated. A typical measurement range is 0–10 mm.

The second form of variable-inductance transducer has a very similar size and physical appearance to the LVDT, but has a center-tapped single winding, as shown in figure 9.3(b). The two halves of the winding are connected to form two arms of a bridge circuit as shown in figure 9.4, which is excited with an alternating voltage. With the core in the central position, the output from the bridge is zero. Displacements of the core on either side of the null position cause a net output voltage which is approximately proportional to the displacement for small movements of the core. Instruments in this second form are available to cover a wide span of displacement measurements. At the lower end of this span, instruments with a range of 0–2 mm are available, whilst at the top end, instruments with a range of 0–5 m can be obtained.

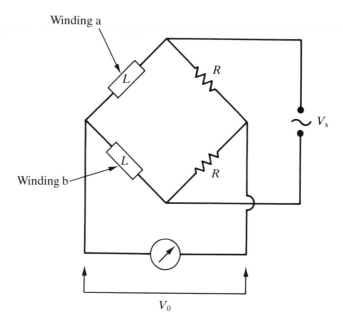

Figure 9.4   Connection of variable-inductance transducer windings in bridge circuit

Both forms of variable-inductance transducer have infinite measurement resolution and a quoted life expectancy of 200 years. Their typical accuracy of ±0.5% of full scale is better than that of the LVDT. In spite of this, their use is very much less common than that of the LVDT.

9.4   Variable-capacitance transducers

Like variable inductance, the principle of variable capacitance is used in displacement-measuring transducers in various ways. The three most common forms of variable-capacitance transducer are shown in figure 9.5. In figure 9.5(a), the capacitor plates are formed by two concentric, hollow, metal cylinders. The displacement to be measured is applied to the inner cylinder which alters the capacitance. The second form, figure 9.5(b), consists of two flat, parallel, metal plates, one of which is fixed and one of which is movable. Displacements to be measured are applied to the movable plate and the capacitance changes as this moves. Both of these first two forms use air as the dielectric medium between the plates. The final form, figure 9.5(c), has two flat, parallel, metal plates with a sheet of solid dielectric material between them. The displacement to be measured causes a capacitance change by moving the dielectric sheet.

(a)

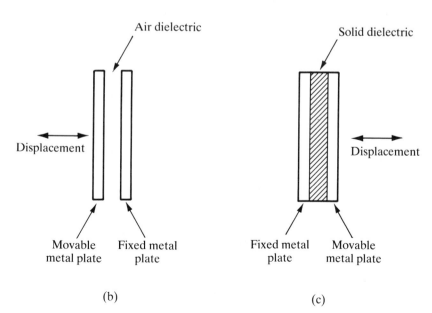

(b)   (c)

Figure 9.5   Variable-capacitance transducers

Accuracies up to ±0.01% are possible with these instruments with measurement resolutions of 1 μm. Individual devices can be selected from manufacturers' ranges which measure displacements as small as 10^{-11} m or as large as 1 m. The fact that such instruments consist only of two simple conducting plates means that it is possible to fabricate devices which are tolerant to a wide range of environmental hazards such as extreme temperatures, radiation and corrosive atmospheres. As there are no contacting moving parts, there is no friction or wear in operation and the life expectancy quoted is 200 years. The major problem with variable-capacitance transducers is their high impedance. This makes them very susceptible to noise and means that the length and position of connecting cables need to be chosen very carefully. In addition, very high-impedance instruments need to be used to measure the value of the capacitance. Because of these difficulties, use of these devices tends to be limited to those few applications where the high accuracy and measurement resolution of the instrument are required.

9.5   Strain gauges

A strain gauge is a device which experiences a change in resistance when it is stretched or strained. It consists physically of a length of resistance wire formed into a zig-zag pattern and mounted onto a flexible backing sheet, as shown in figure 9.6. The wire is nominally of circular cross-section. As strain is applied to the gauge, the shape of the cross-section of the resistance wire distorts, changing the cross-sectional area. As the resistance of the wire per unit length is inversely proportional to the cross-sectional area, there is a consequential change in resistance. The input–output relationship of a strain gauge is expressed by the gauge factor, which is defined as the change in resistance (R) for a given value of strain (S), i.e.

$$\text{Gauge factor} = \frac{\Delta R}{\Delta S}$$

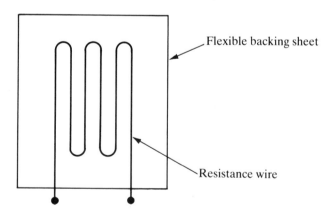

Figure 9.6   The strain gauge

Both metal alloy and semiconductor elements are used in strain gauges. Of the two, semiconductor elements have a much superior gauge factor (up to 50× better than metal types) but are more expensive. Also, whilst metal-alloy gauges have an almost zero temperature coefficient, semiconductor types have a relatively high temperature coefficient.

In use, strain gauges are bonded to the object whose displacement is to be measured. The process of bonding presents a certain amount of difficulty, particularly for semiconductor types. The resistance of the gauge is usually measured by a d.c. bridge circuit and the displacement is inferred from the bridge output measured. As the resistance change in a strain gauge is small, the bridge output voltage is also small and amplification has to be carried out. This adds to the cost of using strain gauges for displacement measurement.

A very common application of strain gauges is in measuring the very small displacements which occur in the diaphragms of pressure-measuring instruments. The strain gauge has a typical range of measurement of 0–50 μm and so is very suitable for this task. Measurement accuracies of ±0.15% of full-scale reading are achievable and the quoted life expectancy is usually 3 million reversals.

The very small range of displacement measurement can be extended grossly by the scheme illustrated in figure 9.7. In this, the displacement to be measured is applied to a wedge fixed between two beams carrying strain gauges. As the wedge is displaced downward, the beams are forced apart and strained, causing an output reading on the strain gauges. Using this method, displacements up to about 50 mm can be measured.

Figure 9.7   Strain gauges measuring large displacements

9.6   Nozzle flapper

The nozzle flapper is a displacement transducer which translates displacements into a pressure change. A secondary pressure-measuring device is therefore required within the instrument. The general form of a nozzle flapper is shown schematically in figure 9.8. Fluid at a known supply pressure, P_s, flows through a fixed restriction and then through a variable restriction formed by the gap, x, between the end of the main vessel and the flapper plate. The body whose displacement is being measured is connected physically to the flapper plate. The output measurement of the instrument is the pressure P_0 in the chamber shown in figure 9.8, and this is almost proportional to x over a limited range of movement of the flapper plate. The instrument typically has a first-order response characteristic. Air is very commonly used as the working fluid and this gives the instrument a time constant of about 0.1 s. The instrument has extremely high sensitivity but its range of measurement is quite small. A typical measurement range is ± 0.05 mm with a measurement resolution of ± 0.01 μm. One very common application of nozzle flappers is measuring the displacements within a load cell, which are typically very small.

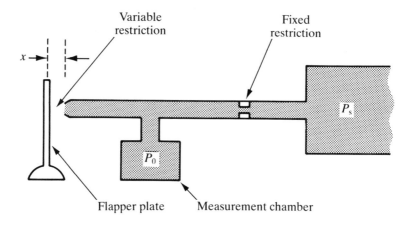

Figure 9.8   The nozzle flapper

9.7   Piezoelectric transducers

The piezoelectric transducer is effectively a force-measuring device which is used in many instruments measuring force, or the force-related quantities of pressure and acceleration. It is included within this discussion of linear displacement transducers because its mode of operation is to generate an e.m.f. proportional to the distance that it is compressed by.

   The device is manufactured from a crystal, which can be either a natural material such as quartz or a synthetic material such as lithium sulfate. The crystal is mechanically stiff, i.e. a large force is required to compress it, and consequently piezoelectric transducers can only be used to measure the displacement of mechanical systems which are stiff enough themselves to be unaffected by the stiffness of the crystal. When the crystal is compressed, a charge is generated on the surface which is measured as the output voltage. As is normal with any induced charge, the charge leaks away over a period of time. Consequently, the output voltage–time characteristic is as shown in figure 9.9. Because of this, piezoelectric transducers are not suitable for measuring static or slowly varying displacements, even though the time constant of the charge decay process can be lengthened by adding a shunt capacitor across the device.

As a displacement-measuring device, the piezoelectric transducer has a very high sensitivity, about 1000× better then the strain gauge. Its typical accuracy is ±1% of full-scale reading and its life expectancy is 3 million reversals.

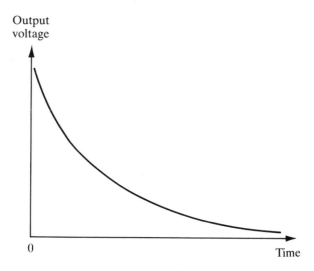

Figure 9.9   Voltage–time characteristic of piezoelectric transducer following step displacement

9.8   Other methods of measuring small displacements

Apart from the methods outlined above, several other techniques for measuring small translational displacements exist, as discussed below. Some of these involve special instruments which have a very limited sphere of application, for instance in measuring machine tool displacements. Others are very recent developments which may potentially gain wide use in the future but have few applications at present.

Linear inductosyn

The linear inductosyn is an extremely accurate instrument which is widely used for axis measurement and control within machine tools. Typical measurement resolution is 2.5 μm.

The instrument consists of two magnetically coupled parts which are separated by an air gap, typically 0.125 mm wide, as shown in figure 9.10. One part, the track, is attached to the axis along which displacements are to be measured. This would

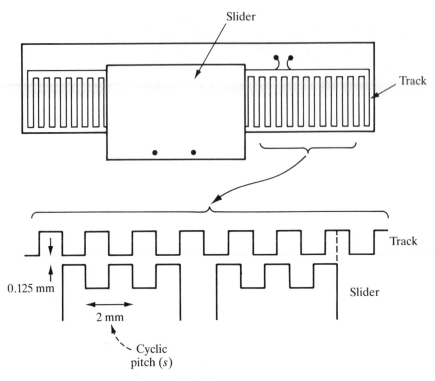

Figure 9.10   The linear inductosyn

generally be the bed of a machine tool. The other part, the slider, is attached to the body which is to be measured or positioned. This would usually be a cutting tool.

The track, which may be several meters long, consists of a fine metal wire formed into the pattern of a continuous rectangular waveform and deposited onto a glass base. The typical pitch (cycle length), s, of the pattern is 2 mm, and this extends over the full length of the track. The slider is usually about 50 mm wide and carries two separate wires formed into continuous rectangular waveforms which are displaced with respect to each other by one-quarter of the cycle pitch, i.e. by 90 electrical degrees.

The wire waveform on the track is excited by an applied voltage given by:

$$V_s = V \sin(\omega t)$$

This excitation causes induced voltages in the slider windings. When the slider is positioned in the null position such that its first winding is aligned with the winding on the track, the output voltages on the two slider windings are given by:

$$V_1 = 0 \qquad V_2 = V \sin(\omega t)$$

For any other position, the slider winding voltages are given by:

$$V_1 = V \sin(\omega t) \sin(2\pi x/s) \qquad V_2 = V \sin(\omega t) \cos(2\pi x/s)$$

where x is the displacement of the slider away from the null position.

Consideration of these equations for the slider-winding outputs shows that the pattern of output voltages repeats every cycle pitch. The instrument can therefore only discriminate displacements of the slider within one cycle pitch of the windings. This means that the typical measurement range of an inductosyn is only 2 mm. This is of no use in normal applications and therefore an additional displacement transducer with coarser resolution but larger measurement range has to be used as well. This coarser measurement is commonly made by translating the linear displacements into rotary motion by suitable gearing which is then measured by a rotational displacement transducer such as a synchro or resolver.

One slight problem with the inductosyn is the relatively low level of coupling between the track and slider windings. Compensation for this is made by using a high-frequency excitation voltage (5–10 kHz is common).

Translation of linear displacements into rotary motion

In some applications, it is inconvenient to measure linear displacements directly, either because there is insufficient space to mount a suitable transducer or because it is inconvenient for other reasons. A suitable solution in such cases is to translate the translational motion into rotational motion by suitable gearing. Any of the rotational displacement transducers discussed in chapter 11 can then be applied.

Integration of output from velocity transducers and accelerometers

If velocity transducers or accelerometers already exist in a system, displacement measurements can be obtained by integration of the output from these instruments. This, however, only gives information about the relative position with respect to some arbitrary starting point. It does not yield a measurement of the absolute position of a body in space unless all motions away from a fixed starting point are recorded.

Laser interferometer

This recently developed instrument is shown in figure 9.11. In this particular design, a dual-frequency helium–neon (He–Ne) laser is used which gives an output pair of light waves at a nominal frequency of 5×10^{14} Hz. The two waves differ in frequency by 2×10^6 Hz and are of opposite polarization. This dual-frequency output waveform is split into measurement and reference beams by the first beam splitter.

The reference beam is sensed by the polarizer and photodetector, A, which converts both waves in the light to the same polarization. The two waves interfere constructively and destructively alternately, producing light–dark flicker at a frequency of 2×10^6 Hz. This excites a 2 MHz electrical signal in the photodetector.

The measurement beam is separated into the two component frequencies by a polarizing beam splitter. Light of the first frequency, f_1, is reflected by a fixed reflecting cube into a photodetector and polarizer, B. Light of the second frequency, f_2, is reflected by a movable reflecting cube and also enters B. The displacement to be measured is applied to the movable cube. With the movable cube in the null position, the light waves entering B produce an electrical signal output at a frequency of

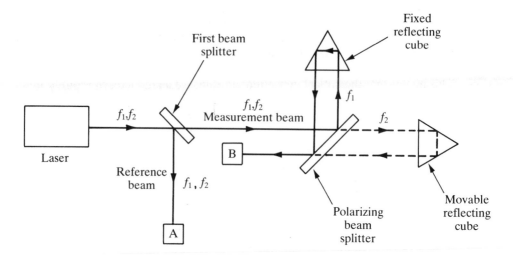

First beam
splitter

Fixed
reflecting
cube

f_1, f_2

f_1, f_2
Measurement beam

f_1

f_2

Laser

Reference
beam

f_1, f_2

B

A

Polarizing
beam
splitter

Movable
reflecting
cube

Figure 9.11   The laser interferometer

2 MHz, which is the same frequency as the reference signal output from A. Any displacement of the movable cube causes a Doppler shift in the frequency f_2 and changes the output from B. The frequency of the output signal from B varies between 0.5 and 3.5 MHz according to the speed and direction of movement of the movable cube.

The outputs from A and B are amplified and subtracted. The resultant signal is fed to a counter whose output indicates the magnitude of the displacement in the movable cube and whose rate of change indicates the velocity of motion.

This technique is beginning to be used in applications requiring high-accuracy measurement such as machine tool control. Such systems can measure to an accuracy of a few parts per million over ranges up to 2 m long. They are therefore an attractive alternative to the inductosyn, in having both high measurement resolution and a large measurement range within one instrument.

Fotonic sensor

The Fotonic sensor is one of many instruments developed recently which make use of fiber optic techniques. It consists of a light source, a light detector, a fiber optic light transmission system and a plate which moves with the body whose displacement is being measured, as shown on figure 9.12. Light from the outward fiber optic cable travels across the air gap to the plate and some of it is reflected back into the return fiber-optic cable. The amount of light reflected back from the plate is a function of the air gap length, x, and hence of the plate displacement. Measurement of the intensity of the light carried back along the return cable to the light detector allows the displacement of the plate to be calculated. A common application of Fotonic sensors is measuring diaphragm displacements in pressure sensors.

Non-contacting optical sensor

Figure 9.13 shows an optical technique which is used to measure small displace-

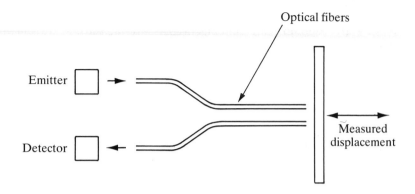

Figure 9.12   The Fotonic sensor

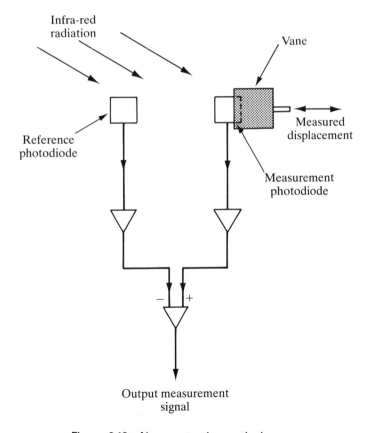

Figure 9.13   Non-contacting optical sensor

ments. The motion to be measured is applied to a vane, whose displacement progressively shades one of a pair of monolithic photodiodes which are exposed to infra-red radiation. A displacement measurement is obtained by comparing the output of the reference (unshaded) photodiode with that of the shaded one. The typical range of measurement is ±0.5 mm with an accuracy of ±0.1% of full scale. Such sensors are used in some intelligent pressure-measuring instruments based on Bourdon tubes or diaphragms as described in section 14.4.

9.9   Measurement of large displacements (range sensors)

One final class of instruments which has not been mentioned so far consists of those designed to measure relatively large translational displacements. Most of these are known as range sensors and measure the motion of a body with respect to some fixed datum point.

Rotary potentiometer and spring-loaded drum

One scheme for measuring large displacements which are beyond the measurement range of common displacement transducers is shown in figure 9.14. This consists of a steel wire attached to the body whose displacement is being measured; the wire passes round a pulley and on to a spring-loaded drum whose rotation is measured by a rotary potentiometer. A multi-turn potentiometer is usually required for this to give an adequate measurement resolution.

Range sensors

Range sensors provide a well used technique of measuring the translational displace-

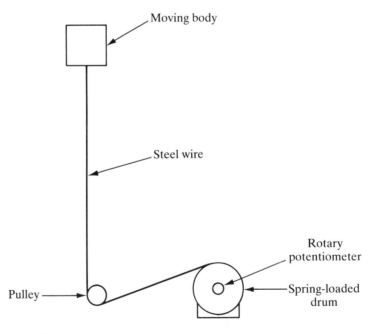

Figure 9.14   System for measuring large displacements

ment of a body with respect to some fixed boundary. The common feature of all range-sensing systems is an energy source, an energy detector and an electronic means of timing the time of flight of the energy between the source and detector. The form of energy used is either ultrasonic or light. In some systems, both energy source and detector are fixed on the moving body and operation depends on the energy being reflected back from the fixed boundary, as in figure 9.15(a). In other systems, the energy source is attached to the moving body and the energy detector is located within the fixed boundary, as shown in figure 9.15(b).

In ultrasonic systems, the energy is transmitted from the source in high-frequency bursts. A frequency of at least 20 kHz is usual and 40 kHz is common for measuring distances up to 5 m. By measuring the time of flight of the energy, the distance of the body from the fixed boundary can be calculated, using the fact that the speed of sound in air is 340 m/s. Because of difficulties in measuring the time of flight with sufficient accuracy, ultrasonic systems are not suitable for measuring distances of less than about 300 mm. Measurement resolution is limited by the wavelength of the ultrasonic energy and can be improved by operating at higher frequencies. At higher frequencies, however, attenuation of the magnitude of the ultrasonic wave as it passes through air becomes significant. Therefore, only low frequencies are suitable if large distances are to be measured. The typical accuracy of ultrasonic range-finding systems is ±0.5% of full scale.

Optical range-finding systems generally use a laser light source. The speed of

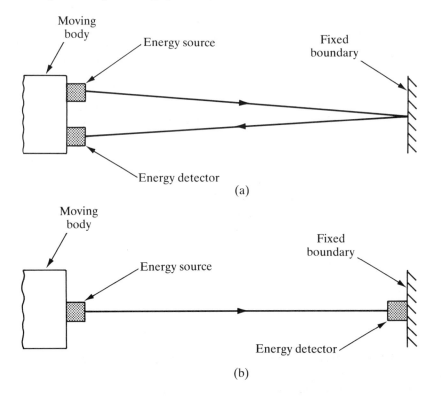

(a)

(b)

Figure 9.15   Range sensors

light in air is about 3×10^8 m/s, so that light takes only a few nanoseconds to travel a meter. In consequence, such systems are only suitable for measuring very large displacements where the time of flight is long enough to be measured with reasonable accuracy.

9.10 Proximity sensors

For the sake of completeness, it is proper to conclude this chapter on translational displacement transducers with consideration of proximity sensors. Proximity detectors provide information on the displacement of a body with respect to some boundary, but only insofar as to say whether the body is less than or greater than a certain distance away from the boundary. The output of a proximity sensor is thus binary in nature: the body is or is not close to the boundary.

Like range sensors, proximity detectors make use of an energy source and detector. The detector is a device whose output changes between two states when the magnitude of the incident reflected energy exceeds a certain threshold level. A common form of proximity sensor uses an infra-red light-emitting diode (LED) source and a phototransistor. Light triggers the transistor into a conducting state when the LED is within a certain distance from a reflective boundary and the reflected light exceeds a threshold level. This system is physically small, occupying a volume of only a few cubic centimeters. If even this small volume is obtrusive, then fiber optic cables can be used to transmit light from a remotely mounted LED and phototransistor. The threshold displacement detected by optical proximity sensors can be varied between 0 and 2 m.

Another form of proximity sensor uses the principle of varying inductance. Such devices are particularly suitable for operation in aggressive environmental conditions and they can be made vibration- and shock-resistant by vacuum encapsulation techniques. The sensor contains a high-frequency oscillator whose output is demodulated and fed via a trigger to an amplifier output stage. The oscillator output radiates through the surface of the sensor and, when the sensor surface becomes close to an electrically or magnetically conductive boundary, the output voltage is reduced because of the interference with the flux paths. At a certain point, the output voltage is reduced sufficiently for the trigger circuit to change state and reduce the amplifier output to zero. Inductive sensors can be adjusted to change state at displacements in the range of 1 to 20 mm.

A third form of proximity sensor uses the capacitive principle. These can operate in similar conditions to inductive types. The threshold level of displacement detected can be varied between 5 and 40 mm.

9.11 Selection of translational measurement transducers

Choice between the various translational motion transducers available for any particular application depends very much on the magnitude of the displacement to be measured.

Displacements larger than 5 m can only be measured by a range sensor, or possibly by the method described in section 9.9 using a wire (figure 9.14). Such methods are also used for displacements in the range between 2 and 5 m, except where the expense of a variable-inductance transducer can be justified.

For measurements within the range of 2 mm to 2 m, the number of suitable instruments grows. Both the relatively cheap potentiometer and the LVDT, which is somewhat more expensive, are commonly used for such measurements. Variable-inductance and -capacitance transducers are also used in some applications. Additionally, strain gauges measuring the strain in two beams forced apart by a wedge (see section 9.5) can measure displacements up to 50 mm. If very high measurement resolution is required, either the linear inductosyn or the laser interferometer is used.

The requirement to measure displacements of less than 2 mm usually occurs as part of an instrument which is measuring some other physical quantity such as pressure, and several types of instrument have evolved to fulfill this task. The LVDT, strain gauges, the Fotonic sensor, variable-capacitance transducers and the non-contacting optical transducer all find application in measuring diaphragm or Bourdon-tube displacements within pressure transducers. Load-cell displacements are also very small, and these are commonly measured by nozzle flapper devices.

Apart from the measurement range required, several other factors influence the choice between different instruments. Severe environmental operating conditions, for example hot, radioactive or corrosive atmospheres, often dictate the use of devices which can be easily protected from these conditions, such as the LVDT, variable-inductance or variable-capacitance instruments. Accessibility and maintenance requirements are also very important factors, as high maintenance costs can quickly destroy the initial advantage that a particular type of instrument may have in terms of low purchase cost.

10

Translational Velocity and Acceleration Measurement

10.1 Measurement of translational velocity

Translational velocity cannot be measured directly and therefore must be calculated indirectly by other means as set out below.

Differentiation of displacement measurements

Differentiation of position measurements obtained from any of the translational displacement transducers described in the last chapter can be used to produce a translational velocity signal. Unfortunately, the process of differentiation always amplifies noise in a measurement system. Therefore, if this method has to be used, a low-noise instrument such as a d.c.-excited carbon-film potentiometer or laser interferometer should be chosen. In the case of potentiometers, a.c. excitation must be avoided because of the problem that harmonics in the power supply would cause.

Integration of the output of an accelerometer

Where an accelerometer is already included within a system, integration of its output can be performed to yield a velocity signal. The process of integration attenuates rather than amplifies measurement noise and this is therefore an acceptable technique.

Conversion to rotational velocity

Conversion from translational to rotational velocity is the final measurement technique open to the system designer and is the one most commonly used. This enables any of the rotational velocity-measuring instruments described in chapter 12 to be applied.

10.2 Measurement of acceleration

The only class of device available for measuring acceleration is the accelerometer.

These are available in a wide variety of types and ranges designed to meet particular measurement requirements. They have a frequency response between zero and a high value and have a form of output which can be readily integrated to give displacement and velocity measurements. The frequency response of accelerometers can be improved by altering the level of damping in the instrument. Such adjustment must be done carefully, however, because frequency response improvements are only achieved at the expense of degrading the measurement sensitivity. Besides their use for general-purpose motion measurement, accelerometers are widely used to measure mechanical shocks and vibrations.

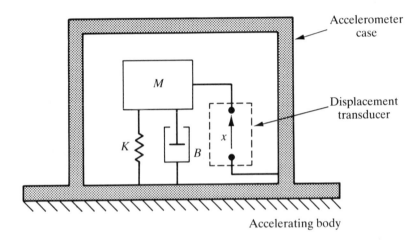

Figure 10.1   Structure of an accelerometer

The structure of most forms of accelerometer is shown in figure 10.1. The accelerometer is rigidly fastened to the body undergoing acceleration. Any acceleration of the body causes a force, F_a, on the mass, M, given by:

$$F_a = M\ddot{x}$$

This force is opposed by the restraining effect, F_s, of a spring with spring constant K, and the net result is that the mass is displaced by a distance x from its starting position such that:

$$F_s = Kx$$

In the steady state, when the mass inside is accelerating at the same rate as the case of the accelerometer, $F_a = F_s$ and so:

$$Kx = M\ddot{x} \quad \text{or} \quad \ddot{x} = (Kx)/M \tag{10.1}$$

This is the equation of motion of a second-order system, and, in the absence of damping, the output of the accelerometer would consist of non-decaying oscillations. A damper is therefore included within the instrument, which produces a damping force, F_d, proportional to the velocity of the mass M given by:

$$Fd = B\dot{x}$$

This modifies the previous equation of motion (10.1) to the following:

$$Kx + b\dot{x} = M\ddot{x} \qquad (10.2)$$

One important characteristic of accelerometers is their sensitivity to accelerations at right angles to the sensing axis (the direction along which the instrument is designed to measure acceleration). This is defined as the cross-sensitivity and is specified in terms of the output, expressed as a percentage of the full-scale output, when an acceleration of some specified magnitude (e.g. $30g$) is applied at 90° to the sensing axis.

The acceleration reading is obtained from the instrument by measurement of the displacement of the mass within the accelerometer. Many different displacement-measuring techniques are used in the various types of accelerometer which are commercially available. Different types of accelerometer also vary in terms of the type of spring element and form of damping used.

Resistive potentiometers are one such displacement-measuring instrument used in accelerometers. These are used mainly for measuring slowly varying accelerations and low-frequency vibrations in the range of 0–$50g$. The measurement resolution obtainable is about 1 in 400 and typical values of cross-sensitivity are $\pm1\%$. Accuracy is about $\pm1\%$ and life expectancy is quoted at 2 million reversals. A typical size and weight are 125 cm^3 and 500 g.

Strain gauges are also used in accelerometers for measuring accelerations up to $200g$. These serve as the spring element as well as measuring mass displacement, thus simplifying the instrument's construction. Their typical characteristics are a resolution of 1 in 1000, accuracy of $\pm1\%$ and cross-sensitivity of 2%. They have a major advantage over potentiometer-based accelerometers in terms of their much smaller size and weight (3 cm^3 and 25 g).

Another displacement transducer found in accelerometers is the LVDT. This class can measure accelerations up to $700g$ with an accuracy of $\pm1\%$ of full scale. They are of a similar physical size to potentiometer-based instruments but are lighter in weight (100 g).

Accelerometers based on variable-inductance displacement-measuring devices have extremely good characteristics and are suitable for measuring accelerations up to $40g$. Typical specifications of such instruments are accuracy $\pm0.25\%$ of full scale, resolution 1 in 10 000 and cross-sensitivity 0.5%. Their physical size and weight are similar to potentiometer-based devices. Such instruments are often found in oil exploration applications measuring vibrations reflected from underground rock strata.

The other common displacement transducer used in accelerometers is the piezoelectric type. The major advantage of using piezoelectric crystals is that they also act as the spring and damper within the instrument. In consequence, the device is quite small (15 cm^3) and very light (50 g), but because of the nature of piezoelectric crystal operation, such instruments are not suitable for measuring constant or slowly time-varying accelerations. As the electrical impedance of a piezoelectric crystal is itself high, the output voltage must be measured with a very high-impedance instrument to avoid loading effects. Many recent piezoelectric-crystal-based accelerometers incorporate a high-impedance charge amplifier within the body of the instrument. This simplifies the signal conditioning requirements external to the accelerometer but can lead to problems in certain operational environments because these internal electronics are exposed to the same environmental hazards as the rest of the accelerometer. Typical measurement resolution of this class of accelerometer is 0.1% of full scale with an accuracy of $\pm 1\%$. Individual instruments are available to cover a wide range of measurements from $0.03g$ full scale up to $1000g$ full scale.

In choosing between the different types of accelerometer available for any particular situation, several factors need to be considered. Of major importance is the mass of the instrument. This should be very much less than that of the body whose motion is being measured, in order to avoid loading effects which affect the accuracy of the readings obtained. In this respect, strain-gauge-based instruments are best.

Vibration measurement

Vibrations are very commonly encountered in machinery operation, and therefore measurement of the accelerations associated with such vibrations is extremely important in industrial environments. The peak accelerations involved in such vibrations can be of $100g$ or greater in magnitude, whilst both the frequency of oscillation and the magnitude of displacements from the equilibrium position in vibrations have a tendency to vary randomly.

Vibrations can be expressed mathematically as:

$$X = X_0 \sin(\omega t) \qquad (10.3)$$

where X is the displacement from the equilibrium position at any general point in time, X_0 is the peak displacement from the equilibrium position, and ω is the angular frequency of the oscillations. By differentiating equation (10.3) with respect to time, an expression for the velocity v of the vibrating body at any general point in time is obtained as:

$$v = -\omega X_0 \cos(\omega t) \qquad (10.4)$$

Differentiating equation (10.4) again with respect to time, we obtain an expression for the acceleration α of the body at any general point in time as:

$$\alpha = -\omega^2 X_0 \sin(\omega t) \qquad (10.5)$$

Inspection of equation (10.5) shows that the peak acceleration is given by:

$$\alpha_{\text{peak}} = \omega^2 X_0 \qquad (10.6)$$

This square-law relationship between peak acceleration and oscillation frequency is the reason why high values of acceleration occur during relatively low-frequency oscillations. For example, an oscillation at 10 Hz produces peak accelerations of $2g$.

The frequency response of accelerometers is particularly important in vibration measurement in view of the inherently high-frequency characteristics of the measurement situation. The bandwidth of both potentiometer-based accelerometers and accelerometers using variable-inductance-type displacement transducers goes up to 25 Hz only. Accelerometers including either the LVDT or strain gauges can measure frequencies up to 150 Hz and the latest instruments using piezoresistive strain gauges have bandwidths up to 2 kHz. Finally, inclusion of piezoelectric-crystal displacement transducers yields an instrument with a bandwidth which can be as high as 7 kHz.

Example
A pipe carrying a fluid vibrates at a frequency of 50 Hz with displacements of 8 mm from the equilibrium position. Calculate the peak acceleration.

Solution
From equation (10.6):

$$\alpha_{peak} = \omega^2 X_0 = (2\pi \times 50)^2 \times (0.008) = 789.6 \text{ m/s}^2$$

Using the fact that the acceleration due to gravity, g, is 9.81 m/s^2, this answer can be expressed alternately as:

$$\alpha_{peak} = 789.6/9.81 = 80.5g$$

Shock measurement

Another particular type of motion which occurs commonly in industrial situations is where some moving body is suddenly brought to rest, often because of a collision. This is known as shock, and is frequently the result of a body being dropped and hitting the floor. Shocks are characterized by involving very high-magnitude decelerations (e.g. 500g) which last for a very short time. An instrument having a very high-frequency response is required for shock measurement, and for this reason, piezoelectric-crystal-based accelerometers are commonly used.

Example
A body is dropped from a height of 10 m and suffers a shock when it hits the ground. If the duration of the shock is 5 ms, calculate the magnitude of the shock in terms of g.

Solution
The equation of motion for a body falling under gravity gives the following expression for the terminal velocity, v:

$$v = (2gx)^{1/2}$$

where x is the height through which the body falls. Having calculated v, the average deceleration during the collision can be calculated as:

$$\alpha = v/t$$

where t is the duration of the shock. Substituting the appropriate numerical values into these expressions:

$$v = (2 \times 9.81 \times 10)^{1/2} = 14.0 \, \text{m/s} \qquad \alpha = 14.0/0.005 = 2801 \, \text{m/s} = 286g$$

11

Rotational Displacement Measurement

Rotational displacement transducers measure the angular motion of a body about some rotation axis. They are important not only for measuring the rotation of bodies such as shafts, but also as part of systems which measure translational displacement by converting the translational motion to a rotary form. The various devices available for measuring rotational displacements are presented below, and the arguments for choosing a particular form in any given measurement situation are considered at the end of the chapter.

11.1 Circular and helical potentiometers

This class of instrument is the cheapest available for measuring rotational displacements, with the circular type being the cheaper of the two. Their structure is shown in figure 11.1. The mode of operation is exactly the same as for translational motion potentiometers, except that the track is bent round into a circular shape. The output voltage measured at the sliding contact is proportional to the angular displacement of the sliding contact from the starting position, thus satisfying the usual requirement for a linear measured quantity – output reading relationship.

(a)   (b)

Figure 11.1   Rotary motion potentiometers: (a) circular; (b) helical

The circular potentiometer has a track consisting of no more than one full turn, and hence has a maximum angular measurement range of 0–360°, although they are often produced with measurement ranges as small as 0–10°. The measurement resolution afforded by this arrangement is very limited, and is the reason for the existence of multi-turn devices where the resistance element is formed into the shape of a helix. Such devices are produced with up to 60 turns. The accuracy of such devices varies from ±1% of full scale for circular potentiometers up to ±0.002% of full scale for the best helical potentiometers.

11.2 Rotational differential transformer

Special forms of differential transformer are available which measure rotational rather than translational motion. The method of construction and connection of the windings is exactly the same as for the linear variable differential transformer (LVDT), except that a specially shaped core is used which varies the mutual inductance between the windings as it rotates, as shown in figure 11.2. Apart from the difficulty of avoiding some asymmetry between the secondary windings, great care has to be taken in these instruments to machine the core to exactly the right shape.

Figure 11.2   Rotary differential transformer

In consequence, the best accuracy obtainable is only ±1%. This accuracy is only obtained for limited excursions of the core of ±40° away from the null position. For angular displacements of ±60°, the accuracy drops to ±3%, and the instrument is unsuitable for measuring displacements greater than this. Like its linear equivalent, the instrument suffers no wear in operation and therefore has a very long life with almost no maintenance requirements. It can also be modified for operation in harsh environments by enclosing the windings inside a protective enclosure.

11.3 Incremental shaft encoders

Incremental shaft encoders are one of a class of encoder devices which give an output

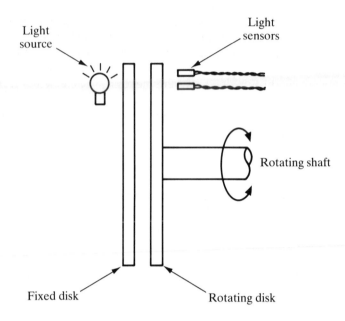

Figure 11.3   The optical incremental shaft encoder

in digital form. They measure the instantaneous angular position of a shaft relative to some arbitrary datum point, but are unable to give any indication about the absolute position of a shaft. The principle of operation is to generate pulses as the shaft whose displacement is being measured rotates. These pulses are counted and the total angular rotation inferred from the pulse count. The pulses are generated either by optical or by magnetic means and are detected by suitable sensors. Of the two, the optical system is considerably cheaper and therefore much more common. Such instruments are very convenient for computer control applications, as the measurement is already in the required digital form and therefore the usual analog-to-digital signal conversion process is avoided.

   An example of an optical incremental shaft encoder is shown in figure 11.3. It can be seen that the instrument consists of a pair of disks, one of which is fixed and one of which rotates with the body whose angular displacement is being measured. Each disk is basically opaque but has a pattern of windows cut into it. The fixed disk has only one window and the light source is aligned with this so that the light shines through all the time. The second disk has two tracks of windows cut into it which are equidistantly spaced around the disk, as shown in figure 11.4. Two light detectors are positioned beyond the second disk so that one is aligned with each track of windows. As the second disk rotates, light alternately enters and does not enter the detectors, as windows and then opaque regions of the disk pass in front of them. These pulses of light are fed to a counter, with the final count after motion has ceased corresponding to the angular position of the moving body relative to the starting position. The primary information about the magnitude of rotation is obtained by the detector aligned with the outer track of windows. The pulse count obtained from this gives no information about the direction of rotation, however. Direction information is

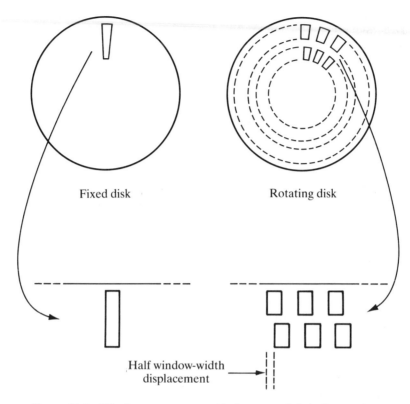

Fixed disk   Rotating disk

,Half window-width
displacement

Figure 11.4   Window arrangement in incremental shaft encoder

provided by the second, inner track of windows, which have an angular displacement
with respect to the outer set of windows of half a window-width. The pulses from the
detector aligned with the inner track of windows therefore lag or lead the primary set
of pulses according to the direction of rotation.

The maximum measurement resolution obtainable is limited by the number of
windows which can be machined onto a disk. The maximum number of windows per
track for a 150 mm diameter disk is 5000, which gives a basic angular measurement
resolution of 1 in 5000. By using more sophisticated circuits which increment the
count on both the rising and falling edges of the pulses through the outer track of
windows, it is possible to double the resolution to a maximum of 1 in 10 000. At the
expense of even greater complexity in the counting circuit, it is possible also to
include the pulses from the inner track of windows in the count, so giving a maximum
measurement resolution of 1 in 20 000.

Optical incremental shaft encoders are a popular instrument for measuring
relative angular displacements and are generally reliable in operation. Problems of
noise in the system giving false counts can sometimes cause difficulties, although this
can usually be eliminated by squaring the output from the light detectors.

Such instruments are found in many applications where rotational motion has
to be measured. Incremental shaft encoders are also commonly used in circumstances
where a translational displacement has been transformed to a rotational one by suitable gearing. One example of this practice is in measuring the translational motions

in numerically controlled (NC) drilling machines. Typical gearing used for this would give one revolution per millimeter of translational displacement. By using an incremental shaft encoder with 1000 windows per track in such an arrangement, a measurement resolution of 1 μm is obtained.

11.4 Coded disk shaft encoders

Unlike the incremental shaft encoder, which gives a digital output in the form of pulses that have to be counted, the digital shaft encoder has an output in the form of a binary number of several digits which provides an absolute measurement of shaft position. This is particularly useful for computer control applications. Three different forms of digital shaft encoder exist using respectively optical, electrical and magnetic energy systems.

Optical digital shaft encoder

The optical digital shaft encoder is the cheapest form available of this class of instrument and is the one used most commonly. It is found in a variety of applications, and one where it is particularly popular is in measuring the position of rotational joints in robot manipulators.

The instrument is similar in physical appearance to the incremental shaft encoder, in having a pair of disks (one movable and one fixed) with a light source on one side and light detectors on the other side, as shown in figure 11.5. The fixed disk

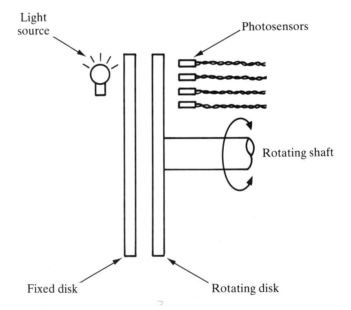

Figure 11.5   The coded disk shaft encoder

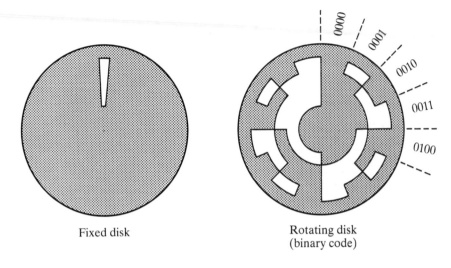

Fixed disk   Rotating disk
                    (binary code)

Figure 11.6   Window arrangement for coded disk shaft encoder

has a single window as in the incremental shaft encoder, and the principal way in which the device differs from the incremental shaft encoder is in the design of the windows on the movable disk, as shown in figure 11.6. These are cut in four or more tracks instead of two and are arranged in sectors as well as tracks. An energy detector is aligned with each track, and these give an output of '1' when energy is detected and an output of '0' otherwise. The measurement resolution obtainable depends on the number of tracks used. For a four-track version, the resolution is 1 in 16, with progressively higher measurement resolution being attained as the number of tracks is increased. These binary outputs from the detectors are combined together to give a binary number of several digits. The number of digits corresponds to the number of tracks on the disk, which in the example shown in figure 11.6 is four. The pattern of windows in each sector is cut such that, as that particular sector passes across the window in the fixed disk, the four energy detector outputs combine to give a unique binary number. In the binary-coded example shown in figure 11.6, the binary number output increments by one as each sector in the rotating disk passes in turn across the window in the fixed disk. Thus the output from sector 1 is 0001, from sector 2 is 0010, from sector 3 is 0011, etc.

Whilst this arrangement is perfectly adequate in theory, serious problems can arise in practice due to the manufacturing difficulty of machining the windows of the movable disk such that the edges of the windows in each track are aligned with each other exactly. Any misalignment means that, as the disk turns across the boundary between one sector and the next, the outputs from each track will switch at slightly different instants of time, and therefore the binary number output will be incorrect over small angular ranges corresponding to the sector boundaries. The worst error can occur at the boundary between sectors 7 and 8, where the output is switching from 0111 to 1000. If the energy sensor corresponding to the first digit switches before the others, then for a very small angular range of movement the output will be 1111, indicating that sector 15 is aligned with the fixed disk rather than sector 7 or 8. This represents an error of 100% in the indicated angular position.

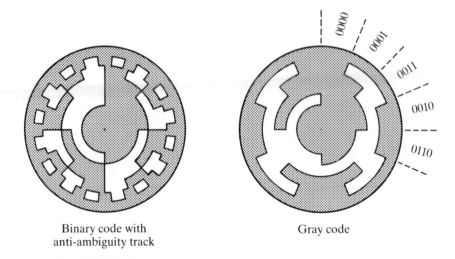

Binary code with
anti-ambiguity track

Gray code

Figure 11.7   Modified window arrangements for the rotating disk

There are two ways used in practice to overcome this difficulty, which both involve an alteration to the manner in which windows are machined on the movable disk, as shown in figure 11.7. The first of these methods adds an extra outer track on the disk, known as an anti-ambiguity track, which consists of small windows which span a small angular range on either side of each sector boundary of the main track system. When energy sensors associated with this extra track sense energy, this is used to signify that the disk is aligned on a sector boundary and the output is unreliable.

The second method is somewhat simpler and cheaper because it avoids the expense of machining the extra anti-ambiguity track. It does this by using a special code known as the Gray code to cut the tracks in each sector on the movable disk. The Gray code is a special binary representation, where only one binary digit changes in moving from one decimal number representation to the next, i.e. from one sector to the next in the digital shaft encoder. The code is illustrated in table 11.1.

It is possible to manufacture optical digital shaft encoders with up to 21 tracks. Such devices have a measurement resolution of 1 part in 10^6 (about one second of arc).

Contacting (electrical) digital shaft encoder

The contacting digital shaft encoder consists of only one disk which rotates with the body whose displacement is being measured. The disk has conducting and non-conducting segments rather than the transparent and opaque areas found on the movable disk of the optical form of instrument, but these are arranged in an identical pattern of sectors and tracks. The disk is charged to a low potential by an electrical brush in contact with one side of the disk, and a set of brushes on the other side of the disk measure the potential in each track. The output of each detector brush is interpreted as a binary value of '1' or '0' according to whether the track in that particular segment is conducting or not and hence whether a voltage is sensed or not. As for the case of the optical form of instrument, these outputs are combined together to give a multi-bit binary number.

Table 11.1   The Gray code

Decimal	Binary	Gray
0	0000	0000
1	0001	0001
2	0010	0011
3	0011	0010
4	0100	0110
5	0101	0111
6	0110	0101
7	0111	0100
8	1000	1100
9	1001	1101
10	1010	1111
11	1011	1110
12	1100	1010
13	1101	1011
14	1110	1001
15	1111	1000

Contacting digital shaft encoders have a similar cost to the equivalent optical instruments and have operational advantages in severe environmental conditions of high temperature or mechanical shock. They suffer from the usual problem of output ambiguity at the sector boundaries but this problem is overcome by the same methods as used in optical instruments.

A serious problem in the application of contacting digital shaft encoders arises from their use of brushes. These introduce friction into the measurement system, and the combination of dirt and brush wear causes contact problems. Consequently, problems of intermittent output can occur, and such instruments generally have limited reliability and a high maintenance cost. Measurement resolution is also limited because of the lower limit on the minimum physical size of the contact brushes. The maximum number of tracks possible is 10, which limits the resolution to 1 part in 1000. Thus contacting digital shaft encoders are only used where the environmental conditions are too severe for optical instruments.

Magnetic digital shaft encoder

Magnetic digital shaft encoders consist of a single rotatable disk, as in the contacting form of encoder discussed in the previous section. The pattern of sectors and tracks consists of magnetically conducting and non-conducting segments, and the sensors aligned with each track consist of small toroidal magnets. Each of these sensors has a coil wound on it which has a high or low voltage induced in it according to the magnetic field close to it. This field is dependent on the magnetic conductivity of that segment of the disk which is closest to the toroid.

These instruments have no moving parts in contact and therefore have a similar reliability to optical devices. Their major advantage over optical equivalents is an ability to operate in very harsh environmental conditions. Unfortunately, the process of manufacturing and accurately aligning the toroidal magnet sensors required makes such instruments very expensive. Their use is therefore limited to a few applications where both high measurement resolution and also operation in harsh environments are required.

11.5 The resolver

The resolver, also known as a synchro-resolver, is an electromechanical device which gives an analog output by transformer action. Physically, resolvers resemble a small a.c. motor and have a diameter ranging from 10 to 100 mm. They have no contacting moving surfaces, and are therefore frictionless and reliable in operation, and have a long life. The best devices give measurement resolutions of 0.1%.

Resolvers have two stator windings, which are mounted at right angles to one another, and a rotor, which can have either one or two windings. As the angular position of the rotor changes, the output voltage changes. The simpler configuration of a resolver with only one winding on the rotor is illustrated in figure 11.8. This exists in two separate forms which are distinguished according to whether the output voltage changes in amplitude or changes in phase as the rotor rotates relative to the stator winding.

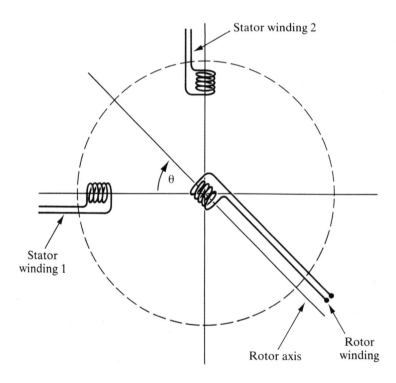

Figure 11.8   Schematic representation of resolver windings

Varying-amplitude output resolver

The stator of this type of resolver is excited with a single-phase sinusoidal voltage of frequency ω, where the amplitudes in the two windings are given by:

$$V_1 = V \sin(\beta) \qquad V_2 = V \cos(\beta)$$

where $V = V_s \sin(\omega t)$. The effect of this is to give a field at an angle of $(\beta + \pi/2)$ relative to stator winding 1. (A full proof of this can be found in Healey (1975).) Suppose that the angle of the rotor winding relative to that of the stator winding is given by θ. Then the magnetic coupling between the windings is a maximum for $\theta = (\beta + \pi/2)$ and a minimum for $\theta = \beta$. The rotor output voltage (see Healey (1975) for proof) is of fixed frequency and varying amplitude given by:

$$V_0 = KV_s \sin(\beta-\theta) \sin(\omega t)$$

This relationship between shaft angle position and output voltage is non-linear, but approximate linearity is obtained for small angular motions where $|(\beta-\theta)| < 15°$.

Varying-phase output resolver

This is a less common form of resolver but is used in a few applications. The stator windings are excited with a two-phase sinusoidal voltage of frequency ω, and the instantaneous voltage amplitudes in the two windings are given by:

$$V_1 = V_s \sin(\omega t) \qquad V_2 = V_s \sin(\omega t + \pi/2) = V_s \cos(\omega t)$$

The net output voltage in the rotor winding is the sum of the voltages induced due to each rotor winding. This is given by:

$$V_0 = KV_s \sin(\omega t) \cos(\theta) + KV_s \cos(\omega t) \cos(\pi/2 - \theta)$$

$$= KV_s[\sin(\omega t) \cos(\theta) + cos(\omega t) \sin(\theta)]$$

$$= KV_s \sin(\omega t + \theta)$$

This represents a linear relationship between shaft angle and the phaseshift of the rotor output relative to the stator excitation voltage. The accuracy of shaft rotation measurement depends on the accuracy with which the phaseshift can be measured. This is improved by increasing the excitation frequency, ω, but this also increases magnetizing losses. Consequently, a compromise excitation frequency of about 400 Hz is used. The conversion of the phaseshift into a varying-amplitude output signal requires relatively complex and expensive electronics. It is mainly for this reason of cost that this type of resolver is not very common.

11.6 The synchro

Like the resolver, the synchro is a motor-like, electromechanical device with an analog output. Apart from having three stator windings instead of two, the instrument is similar in appearance and operation to the resolver and has the same range of physical dimensions. Also, like the resolver, the rotor can have either one or two windings. The rotor usually has a dumbbell shape.

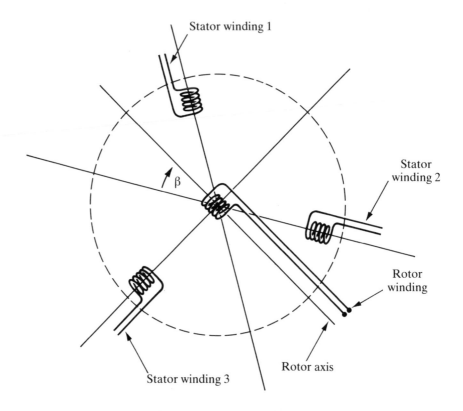

Figure 11.9   Schematic representation of synchro windings

Synchros have been in use for many years for the measurement of angular positions, especially in military applications, and achieve similar levels of accuracy and measurement resolution to digital encoders. One common application is axis measurement in machine tools, where the translational motion of the tool is translated into a rotational displacement by suitable gearing. Synchros are tolerant to high temperatures, high humidity, shock and vibration and are therefore suitable for operation in such harsh environmental conditions. Some maintenance problems are associated with the slip ring and brush system used to supply power to the rotor. However, the only major source of error in the instrument is asymmetry in the windings, and measurement accuracies of ±0.5% are easily achievable.

Figure 11.9 shows the simpler form of synchro with a single rotor winding. If an a.c. excitation voltage is applied to the rotor via slip rings and brushes, this sets up a certain pattern of fluxes and induced voltages in the stator windings by transformer action. For a rotor excitation voltage, V_r, given by:

$$V_r = V \sin(\omega t)$$

the voltages induced in the three stator windings are:

$$V_1 = V \sin(\omega t) \sin(\beta)$$

$$V_2 = V \sin(\omega t) \sin(\beta + 2\pi/3)$$

$$V_3 = V \sin(\omega t) \sin(\beta - 2\pi/3)$$

where β is the angle between the rotor and stator windings.

If the rotor is turned at constant velocity through one full revolution, the voltage waveform induced in each stator winding is as shown in figure 11.10. This has the form of a carrier-modulated waveform in which the carrier frequency corresponds to the excitation frequency, ω. It follows that if the rotor is stopped at any particular angle, β', the peak-to-peak amplitude of the stator voltage is a function of β'. If therefore the stator winding voltage is measured, generally as its root-mean-squared (r.m.s.) value, this indicates the magnitude of the rotor rotation away from the null position. The direction of rotation is determined by the phase difference between the stator voltages, which is indicated by their relative instantaneous magnitudes.

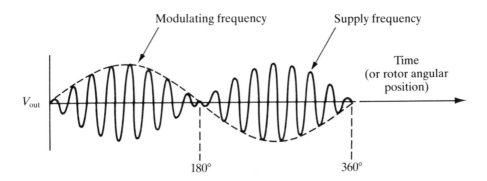

Figure 11.10   Synchro stator voltage waveform

Although a single synchro thus provides a means of measuring angular displacements, it is much more common to find a pair of them used for this purpose. When used in pairs, one member of the pair is known as the synchro transmitter and the other as the synchro transformer, and the two sets of stator windings are connected together, as shown in figure 11.11. Each synchro is of the form shown in figure 11.9, but the rotor of the transformer is fixed for displacement-measuring applications. A sinusoidal excitation voltage is applied to the rotor of the transmitter, setting up a pattern of fluxes and induced voltages in the transmitter stator windings. These voltages are transmitted to the transformer stator windings where a similar

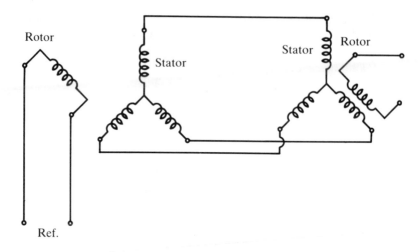

Figure 11.11   Synchro transmitter–transformer pair

flux pattern is established. This in turn causes a sinusoidal voltage to be induced in the fixed transformer rotor winding. For an excitation voltage, $V \sin(\omega t)$, applied to the transmitter rotor, the voltage measured in the transformer rotor is given by:

$$V_0 = V \sin(\omega t) \sin(\theta)$$

where θ is the relative angle between the two rotor windings.

Apart from their use as a displacement transducer, such synchro pairs are commonly used to transmit angular displacement information over some distance, for instance to transmit gyro compass measurements in an aircraft to remote meters. They are also used for load positioning, allowing a load connected to the transformer rotor shaft to be controlled remotely by turning the transmitter rotor. For these applications, the transformer rotor is free to rotate and is also damped to prevent oscillatory motions. In the simplest arrangement, a common sinusoidal excitation voltage is applied to both rotors. If the transmitter rotor is turned, this causes an unbalance in the magnetic flux patterns and results in a torque on the transformer rotor which tends to bring it into line with the transmitter rotor. This torque is typically small for small displacements, and so this technique is only useful if the load torque on the transformer shaft is very small. In other circumstances, it is necessary to incorporate the synchro pair within a servo mechanism, where the output voltage induced in the transformer rotor winding is amplified and applied to a servomotor which drives the transformer rotor shaft until it is aligned with the transmitter shaft.

11.7 The induction potentiometer

These instruments belong to the same class as resolvers and synchros but have only one rotor winding and one stator winding. They are of a similar size and appearance to other devices in the class. A single-phase sinusoidal excitation is applied to the rotor winding and this causes an output voltage in the stator winding through the mutual inductance linking the two windings. The magnitude of this induced stator voltage varies with the rotation of the rotor. The variation of the output with rotation is naturally sinusoidal if the coils are wound such that their field is concentrated at one point, and only small excursions can be made away from the null position if the output relationship is to remain approximately linear. However, if the rotor and stator windings are distributed around the circumference in a special way, an approximately linear relationship for angular displacements of up to ±90° can be obtained.

11.8 The rotary inductosyn

This instrument is similar in operation to the linear inductosyn, except that it measures rotary displacements and has tracks which are arranged radially on two circular disks, as shown in figure 11.12. Typical diameters of the instrument vary between 75 and 300 mm. The larger versions give a measurement resolution of up to 0.05 seconds of arc. Like its linear equivalent, however, the rotary inductosyn has a very small measurement range, and a lower-resolution rotary displacement transducer with a larger measurement range must be used in conjunction with it.

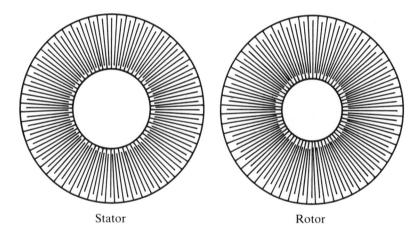

Stator   Rotor

Figure 11.12   The rotary inductosyn

11.9 Gyroscopes

Gyroscopes measure both absolute angular displacement and absolute angular velocity. They consist essentially of a large, motor-driven wheel whose angular momentum is such that the axis of rotation tends to remain fixed in space, thus acting as a reference point. The gyro frame is attached to the body whose motion is to be measured. The output is measured in terms of the angle between the frame and the axis of the spinning wheel. Two different forms of gyroscope are used for measuring angular displacement, the free gyro and the rate-integrating gyro.

The free gyroscope

A free gyroscope is illustrated in figure 11.13. This measures the absolute angular rotation of the body to which its frame is attached about two perpendicular axes. Two alternate methods of driving the wheel are used in different versions of the instrument. One of these is to enclose the wheel in stator-like coils which are excited with a sinusoidal voltage. A voltage is applied to the wheel via slip rings at both ends of the spindle carrying the wheel. The wheel behaves as a rotor and motion is produced by motor action. The other, less common, method is to fix vanes on the wheel which is then driven by directing a jet of air onto the vanes.

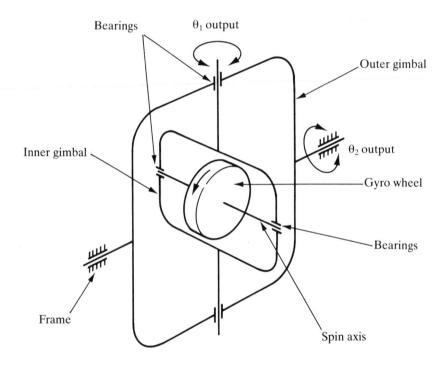

Figure 11.13   The free gyroscope

The instrument can measure angular displacements of up to 10° with a high accuracy. For greater angular displacements, interaction between the measurements on the two perpendicular axes starts to cause a serious loss of accuracy. The physical size of the coils in the motor-action-driven system also limits the measurement range to 10°. For these reasons, this type of gyroscope is only suitable for measuring rotational displacements of up to 10°.

A further operational problem of free gyroscopes is the presence of angular drift (precession) due to bearing friction torque. This has a typical magnitude of 0.5° per minute and means that the instrument can only be used over short time intervals of say 5 min. This time duration can be extended if the angular momentum of the spinning wheel is increased.

A major application of gyroscopes is in inertial navigation systems. Only two free gyros mounted along orthogonal axes are needed to monitor motions in three dimensions, because each instrument measures displacement about two axes. The limited angular range of measurement is not usually a problem in such applications, as control action prevents the error in the direction of motion about any axis ever exceeding 1 or 2°. Precession is a much greater problem, however, and for this reason the rate-integrating gyro is used much more commonly.

Rate-integrating gyroscope

The rate-integrating gyroscope, or 'integrating gyro' as it is commonly known, is illustrated in figure 11.14. It measures angular displacements about a single axis only, and therefore three instruments are required in a typical inertial navigation system. The major advantage of the instrument over the free gyro is the almost total absence of precession, with typical specifications quoting drifts of only 0.01° per hour. The instrument has a first-order type of response given by:

$$\frac{\theta_0}{\theta_i}(D) = \frac{K}{\tau D + 1} \tag{11.1}$$

where $K = H/\beta$, $\tau = M/\beta$, θ_i is the input angle, θ_0 is the output angle, D is the D operator, H is the angular momentum, M is the moment of inertia of the system about the measurement axis and β is the damping coefficient. (The mathematical derivation of this equation is given in Doebelin (1975).)

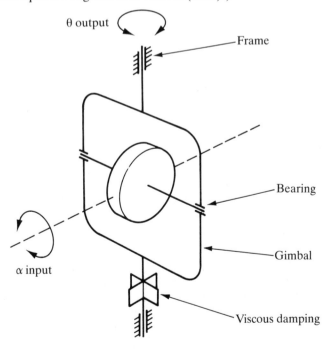

Figure 11.14   The rate-integrating gyroscope

Inspection of equation (11.1) shows that to obtain a high value of measurement sensitivity, K, a high value of H and low value of β are required. A large H is normally obtained by driving the wheel with a hysteresis-type motor revolving at high speeds of up to 24 000 r.p.m. The damping coefficient β can only be reduced so far, however, because a small value of β results in a large value for the system time constant, τ, and an unacceptably low speed of system response. Therefore, the value of β has to be chosen as a compromise between these constraints.

Besides their use as a fixed reference in inertial guidance systems, integrating gyros are also commonly used within aircraft autopilot systems and in military applications such as stabilizing weapon systems in tanks.

Ring laser gyroscope

The ring laser gyroscope is a new type of instrument which is likely to make a major impact on the gyroscope market. The instrument consists of a glass ceramic chamber containing a helium–neon gas mixture in which laser beams are generated. Three mirrors, supported by the ceramic block and mounted in a triangular arrangement, direct a pair of laser beams around the cavity in opposite directions. Any rotation of the ring affects the coherence of the two beams, raising one in frequency and lowering the other. Measurement of the frequency difference allows the angle of rotation to be determined. A more detailed description of the mode of operation is to be found in the journal *IEE News* (1985).

The advantages of this new type of instrument are considerable. The measurement accuracy obtained is substantially better than that afforded by other types of gyro in a similar price range. The device is also considerably smaller physically, which is of considerable benefit in many applications.

11.10 Choice between different instruments

Choice between the various rotational displacement transducers which might be used in any particular measurement situation depends first of all upon whether absolute measurement of angular position is required or whether the measurement of rotation relative to some arbitrary starting point is acceptable. Other factors affecting the choice between instruments are the required measurement range, the resolution of the transducer and the measurement accuracy afforded.

Where only measurement of relative angular position is required, the incremental encoder is a very suitable instrument. The best commercial instruments of this type can measure rotations to a resolution of 1 part in 20 000 of a full revolution, and the measurement range is an infinite number of revolutions. Instruments with such a high measurement resolution are very expensive, but much cheaper versions are available according to what lower level of measurement resolution is acceptable.

All the other instruments presented in this chapter provide an absolute measurement of angular position. The required measurement range is a dominant factor in the choice between these. If this exceeds one full revolution, then the only instrument available is the helical potentiometer. Such devices can measure rotations of up to 60 full turns, but they are expensive because the procedure involved in manufacturing a helical resistance element to a reasonable standard of accuracy is difficult.

For measurements of less than one full revolution, the range of available instruments widens. The cheapest one available is the circular potentiometer, but much better measurement accuracy and resolution is obtained from coded disk encoders. The cheapest of these is the optical form, but certain operating environments necessitate the use of the alternate contacting (electrical) and magnetic versions. All types of coded disk encoder are particularly attractive in computer control schemes, as the output of encoders is in digital form. A varying-phase output resolver is yet another instrument which can measure angular displacements up to one full revolution in magnitude. Unfortunately, these instruments are expensive because of the complicated electronics incorporated to measure the phase variation and convert it to a varying-amplitude output signal, and hence their use is not common.

An even greater range of instruments becomes available as the required measurement range is reduced further. These include the synchro ($\pm90°$), the varying-amplitude output resolver ($\pm90°$), the induction potentiometer ($\pm90°$) and the differential transformer ($\pm40°$). All these instruments have a high reliability and a long service life.

Finally, two further instruments are available for satisfying special measurement requirements, the rotary inductosyn and the gyroscope. The rotary inductosyn is used in applications where very high measurement resolution is required, although the measurement range afforded is extremely small and a coarser-resolution instrument must be used in parallel to extend the measurement range. Gyroscopes, in both the free gyro and rate-integrating gyro forms, are used to measure small angular displacements up to $\pm10°$ in magnitude in inertial navigation systems and similar applications.

12

Rotational Velocity Measurement

The main application of rotational velocity transducers is in speed control systems. They also provide the usual means of measuring translational velocities, which are transformed into rotational motions for measurement purposes by suitable gearing. Many different instruments and techniques are available for measuring rotational velocity as presented below. The factors governing choice between them for particular applications are discussed at the end of the chapter.

12.1 Differentiation of angular displacement measurements

Angular velocity measurements can be obtained by differentiating the output signal from angular displacement transducers. Unfortunately, the process of differentiation amplifies any noise in the measurement signal, and therefore this technique is rarely used.

12.2 Integration of the output from an accelerometer

In measurement systems which already contain an angular acceleration transducer, such as a gyro accelerometer, it is possible to obtain a velocity measurement by integrating the acceleration measurement signal. This produces a signal of acceptable quality, as the process of integration attenuates any measurement noise. The method is of limited value in many measurement situations, however, because the measurement obtained is the average velocity over a period of time, rather than a profile of the instantaneous velocities as motion takes place along a particular path.

12.3 a.c. tachometric generator

The d.c. tachometric generator, or d.c. tachometer as it is generally known, has an output which is approximately proportional to its speed of rotation. Its basic structure is identical to that found in a standard d.c. generator used for producing power,

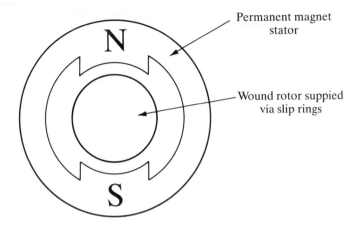

Figure 12.1   The d.c. tachometer

and is shown in figure 12.1. Both permanent-magnet types and separately excited field types are used. However, certain aspects of the design are optimized to improve its accuracy as a speed-measuring instrument. One significant design modification is to reduce the weight of the rotor by constructing the windings on a hollow fiberglass shell. The effect of this is to minimize any loading effect of the instrument on the system being measured.

The d.c. output voltage from the instrument is of a relatively high magnitude, giving a high measurement sensitivity which is typically 5 V per 1000 r.p.m. The direction of rotation is determined by the polarity of the output voltage. A common range of measurement is 0–5000 r.p.m. Maximum non-linearity is usually about ±1% of the full-scale reading.

One problem with these devices which can cause problems under some circumstances is the presence of an a.c. ripple in the output signal. The magnitude of this can be up to 2% of the output d.c. level.

12.4 a.c. tachometric generator

The a.c. tachometric generator, or a.c. tachometer as it is generally known, has an output approximately proportional to rotational speed, as in the d.c. tachogenerator. Its mechanical structure takes the form of a two-phase induction motor, with two stator windings and (usually) a drag-cup rotor, as shown in figure 12.2. One of the stator windings is excited with an a.c. voltage and the measurement signal is taken from the output voltage induced in the second winding. The magnitude of this output voltage is zero when the rotor is stationary, and otherwise proportional to the angular velocity of the rotor. The direction of rotation is determined by the phase of the output voltage, which switches by 180° as the direction reverses. Therefore, both the phase and magnitude of the output voltage have to be measured. A typical range of measurement is 0–4000 r.p.m. with an accuracy of ±0.05% of full-scale reading.

(a) Rotor and stator
winding arrangement

(b) Detail of drag-cup
rotor

Figure 12.1   The a.c. tachometer

   Whilst the form of a.c. tachometer described above is the commonest one, a
second form also exists. This has a squirrel-cage rather than a drag-cup rotor and so
is cheaper. Its structure and mode of operation are otherwise identical, but the
measurement accuracy is reduced.

12.5 Drag-cup tachometric generator

   Like the last two devices, this instrument also has a more common name, the 'drag-

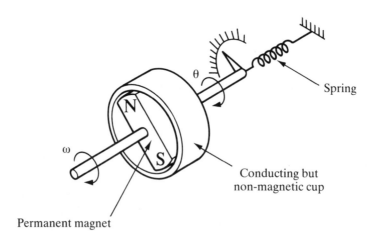

Figure 12.3   The drag-cup tachometer

cup tachometer'. The instrument has a central spindle carrying a permanent magnet which rotates inside a non-magnetic drag-cup consisting of a cylindrical sleeve of electrically conductive material, as shown in figure 12.3. As the spindle and magnet rotate, a voltage is induced which causes circulating eddy currents in the cup. These currents interact with the magnetic field from the permanent magnet and produce a torque. In response, the drag-cup turns until the induced torque is balanced by the torque due to the restraining springs connected to the cup. When equilibrium is reached, the angular displacement of the cup is proportional to the rotational velocity of the central spindle. The instrument has a typical measurement accuracy of ±0.5% and is commonly used in the speedometers of motor vehicles and as a speed indicator for aero engines.

12.6 Variable-reluctance velocity transducer

The form of a variable-reluctance transducer is shown in figure 12.4. It can be seen

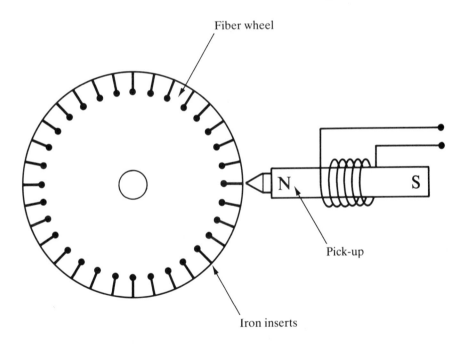

Figure 12.4   Variable-reluctance transducer

that this consists of two parts, a rotating disk connected to the moving body being measured and a pick-up unit. The rotating disk is constructed from a bonded-fiber material into which soft iron poles are inserted at regular intervals around its periphery. The pick-up unit consists of a permanent magnet with a shaped pole piece which carries a wound coil. The distance between the pick-up and the outer perimeter of the disk is around 0.5 mm.

Flux in pick-up

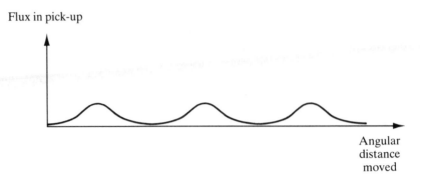

Angular
distance
moved

Figure 12.5   Pattern of flux change with rotation of a
variable-reluctance transducer

As the disk rotates, the soft iron inserts on the disk move in turn past the pick-up unit. As each iron insert moves toward the pole piece, the reluctance of the magnetic circuit increases and hence the flux in the pole piece also increases. Similarly, the flux in the pole piece decreases as each iron insert moves away from the pick-up unit, and the pattern of flux changes with time as shown in figure 12.5. The changing magnetic flux inside the pick-up coil causes a voltage to be induced in the coil whose magnitude is proportional to the rate of change of flux. This voltage is positive whilst the flux is increasing and negative whilst it is decreasing and therefore its variation with time is as shown in figure 12.6.

Pick-up
output
voltage

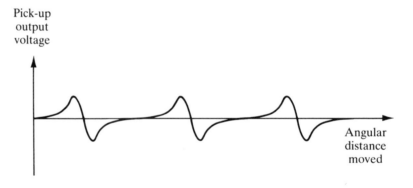

Angular
distance
moved

Figure 12.6   Pattern of induced voltage change with rotation of a
variable-reluctance transducer

The form of this output can be regarded as a sequence of positive and negative pulses whose frequency is proportional to the rotational velocity of the disk. This can be converted into an analog, varying-amplitude, d.c. voltage output by means of a frequency-to-voltage converter circuit connected to the output terminals of the pick-up. However, greater measurement accuracy can be obtained by converting the output waveform into sharp pulses which are counted by an electronic counter. It is normal procedure to produce the pulses at each instant that the induced voltage in the coil changes sign as it passes through zero. This is achieved by electronic means.

The maximum angular velocity which the instrument can measure is limited because of the finite width of the induced pulses. As the velocity increases, the distance between the pulses is reduced, and at a certain velocity the pulses start to overlap. At this point, the pulse counter ceases to be able to distinguish the separate pulses.

A simpler and cheaper form of variable-reluctance transducer uses a ferromagnetic gear wheel in place of a fiber disk. The motion of the tip of each gear tooth toward and away from the pick-up unit causes a similar variation in the flux pattern to that produced by the iron inserts in the fiber disk. The pulses produced by these means are less sharp, however, and consequently the maximum angular velocity measurable is lower.

In either form of variable-reluctance transducer, simply outputting the total pulse count measured over a certain length of time only gives information about the average velocity over that period. Measurement of the actual velocities at the instants of time that each output pulse occurs can be achieved by the scheme shown in figure 12.7. In this circuit, the pulses from the transducer gate the train of pulses from a 1 MHz clock into a counter. Control logic resets the counter and updates the digital output value after receipt of each pulse from the transducer. The measurement resolution of this system is highest when the speed of rotation is low.

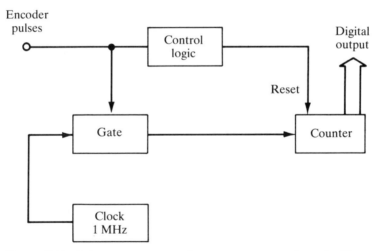

Figure 12.7   Scheme to measure instantaneous angular velocities

12.7 Photoelectric pulse-counting methods

An alternative to the variable-reluctance transducer, but which uses very similar principles to it, is the method where pulses are produced by photoelectric techniques and counted. These pulses are generated by one of the two alternate methods illustrated in figure 12.8. In figure 12.8(a), the pulses are produced as the windows in a slotted disk pass in sequence between a light source and a detector. The alternate

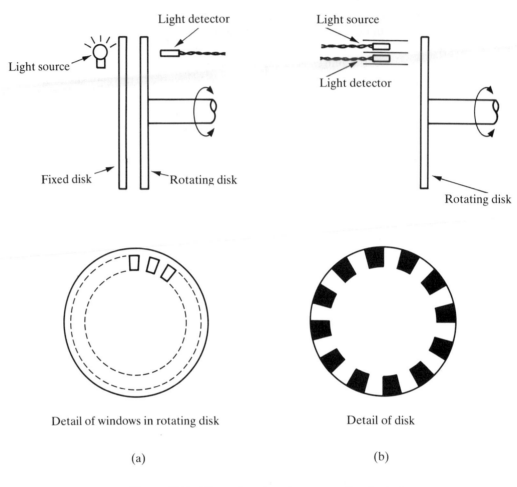

Figure 12.8   Photoelectric pulse generation techniques

form, figure 12.8(b), has both light source and detector mounted on the same side of a reflective disk which has black sectors painted onto it at regular angular intervals. In either case, the pulses are counted by an electronic counter. The frequency of the pulses is proportional to the angular velocity of the body connected to the rotating disk. Pulses generated in this manner are narrower than those generated by magnetic means and so the instrument is capable of measuring higher velocities.

As for the variable-reluctance transducer, a simple pulse count only yields information about the average velocity, but instantaneous velocities can be measured using the scheme illustrated in figure 12.7.

12.8 Stroboscopic methods

The stroboscopic technique of rotational velocity measurement operates on a similar physical principle to the variable-reluctance and photoelectric pulse-counting methods, except that the pulses involved consist of flashes of light generated electronically and whose frequency is adjustable so that it can be matched with the frequency of occurrence of some feature on the rotating body being measured. This feature can either be some naturally occurring one such as the spokes of a wheel or gear teeth, or it can be an artificially created pattern of black and white stripes. In either case, the rotating body appears stationary when the frequencies of the light pulses and body features are in synchronism. Flashing rates up to 25 000 per minute are available from commercial stroboscopes according to the range of velocity measurement required, and the typical measurement accuracy obtained is ±1% of the reading.

Measurement of the flashing rate at which the rotating body appears stationary does not automatically indicate the rotational velocity, because synchronism also occurs when the flashing rate is some integral submultiple of the rotational speed. The practical procedure followed is therefore to adjust the flashing rate until synchronism is obtained at the largest flashing rate possible, R_1. The flashing rate is then carefully decreased until synchronism is again achieved at the next lower flashing rate, R_2. The rotational velocity is then given by:

$$v = \frac{R_1 R_2}{R_1 - R_2}$$

12.9 Mechanical flyball

The mechanical flyball is a velocity-measuring instrument which was first developed many years ago and is still used extensively in speed-governing systems for engines, turbines, etc. As shown in figure 12.9, it consists of a pair of spherical balls pivoted on the rotating shaft. These balls move outward under the influence of centrifugal forces as the rotational velocity of the shaft increases and lift a pointer against the resistance of a spring. The pointer can be arranged to give a visual indication of speed by causing it to move in front of a calibrated scale, or its motion can be converted by a translational displacement transducer into an electrical signal.

In equilibrium, the centrifugal force, F_c, is balanced by the spring force, F_s, where:

$$F_c = K_c \omega^2 \qquad F_s = K_s x$$

and K_c and K_s are constants, ω is the rotational velocity and x is the displacement of the pointer. Thus:

$$K_c \omega^2 = K_s x \qquad \text{or} \qquad \omega = (K_s x / K_c)^{1/2}$$

This is inconvenient because it involves a non-linear relationship between the pointer displacement and the rotational velocity. If this is not accepable, a linear relationship

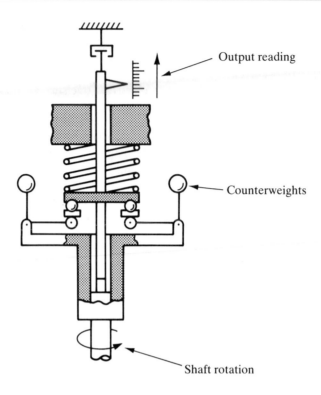

Figure 12.9   The mechanical flyball

can be obtained by using a spring with a non-linear characteristic such that

$$F_s = K'_s x^2$$

This gives:

$$\omega = (K'_s/K_c)^{\frac{1}{2}} x$$

12.10 The rate gyroscope

The rate gyro, illustrated in figure 12.10, has an almost identical construction to the rate-integrating gyro described in the last chapter, and differs only by including a spring system which acts as an additional restraint on the rotational motion of the frame. The instrument measures the absolute angular velocity of a body, and is widely used in generating stabilizing signals within vehicle navigation systems. The typical measurement resolution given by the instrument is 0.01° per second.

The angular velocity, α, of the body is related to the angular deflection of the gyroscope, θ, by the equation:

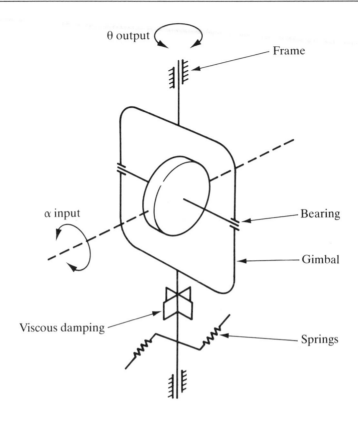

Figure 12.10   The rate gyroscope

$$\frac{\theta}{\alpha}(D) = \frac{HD}{MD^2 + \beta D + K} \tag{12.1}$$

where H is the angular momentum of the spinning wheel, M is the moment of inertia of the system, β is the viscous damping coefficient, K is the spring constant and D is the D operator. (The derivation of this equation is given in Doebelin (1975).)

This relationship (12.1) is a second-order differential equation and therefore we must expect the device to have a response typical of second-order instruments, as discussed in chapter 2. The instrument must therefore be designed carefully so that the output response is neither oscillatory nor too slow in reaching a final reading. To assist in the design process, it is useful to re-express equation (12.1) in the following form:

$$\frac{\theta}{\alpha}(D) = \frac{K'}{D^2/\omega^2 + 2\epsilon D/\omega + 1} \tag{12.2}$$

where $K' = H/K$, $\omega = (K/M)^{1/2}$ and $\epsilon = \beta/2(KM)^{1/2}$. The static sensitivity of the instrument, K', is made as large as possible by using a high-speed motor to spin the wheel and so make H high. Reducing the spring constant, K, further improves the sensitivity but this cannot be reduced too far as it makes the resonant frequency, ω, of the instrument too small. The value of β is chosen such that the damping ratio, ϵ, is close to 0.7.

12.11 Fiber optic gyroscope

This is a relatively new instrument which makes use of fiber-optic technology. Incident light from a source is separated by a beam splitter into a pair of beams a and b, as shown in figure 12.11. These travel in opposite directions around an optic fiber coil and emerge from the coil as the beams marked a' and b'. The beams a' and b' are directed by the beam splitter into an interferometer. Any motion of the coil causes a phase shift between a' and b' which is detected by the interferometer.

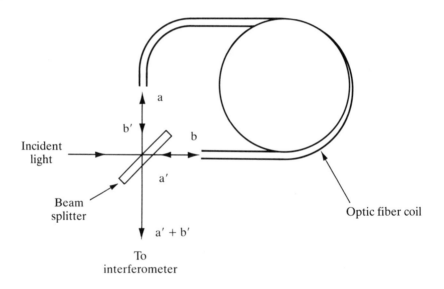

Figure 12.11   The fiber-optic gyroscope

12.12 Choice between different instruments

Choice between different rotational velocity transducers is influenced strongly by whether an analog or digital form of output is required.

Probably the most common form of analog output device used is the d.c. tachometer. This is a relatively simple device which measures speeds up to about 5000 r.p.m. with an accuracy of $\pm 1\%$. Where better accuracy is required within a similar range of speed measurement, a.c. tachometers are used. The squirrel-cage rotor type has an accuracy of $\pm 0.25\%$ and drag-cup rotor types have accuracies up to $\pm 0.05\%$.

Two other devices with an analog output are used in some applications, the drag-cup tachometer and the mechanical flyball. The drag-cup tachometer only has an accuracy of ±5% but is cheap and therefore is very suitable for use in vehicle speedometers. The mechanical flyball has a linear displacement type of output which is commonly connected mechanically by a system of links to the throttle of an engine, thereby fulfilling a speed-governing function.

Digital-output-type instruments are represented by the variable-reluctance transducer, devices using electronic light pulse-counting methods, and the stroboscope. The first two of these are used to measure angular speeds up to about 10 000 r.p.m. and the last one is used to measure speeds up to 25 000 r.p.m.

13

Force and Torque Measurement

Force and torque are closely related entities in the sense that force is a quantity which produces translational motion in a body and torque is a quantity which produces rotational motion in a body. It is therefore appropriate to consider the measurement of both within the same chapter.

13.1 Force measurement

If a force of magnitude F is applied to a body of mass M the body will accelerate at a rate A according to the equation:

$$F = MA$$

This relationship offers two ways of measuring a force, either (i) by measuring the acceleration when the unknown force is applied to a body of known mass, or (ii) by balancing the unknown force against the known gravitational force acting on a standard mass.

Another physical law describing the effect of a force F is the displacement produced when it is applied to a spring or other elastic body according to the equation:

$$F = Kx$$

where K is a constant and x is the displacement produced. Thus method (iii) of force measurement is to measure the displacement produced when the force is applied to an elastic element.

There are also three further general techniques available for measuring force. The first of these, method (iv), converts the task of force measurement into one of pressure measurement by applying the force to a piston in a cylinder containing a constrained fluid. Any of the pressure-measurement transducers discussed in the next chapter can then be applied. Method (v) involves balancing the unknown force against the force of known magnitude created by the interaction between a current-carrying coil and a permanent magnet. Finally, method (vi) uses the variation in resonant frequency of a vibrating wire as it is tensioned by an applied force.

Instruments based on each of these six principles of force measurement will be discussed in the following sections.

Use of accelerometers

The technique of applying a force to a known mass and measuring the acceleration produced can be carried out using any of the types of accelerometer discussed in chapter 10. Unfortunately, the method is of very limited practical value because, in most cases, forces are not free entities but are part of a system (from which they cannot be decoupled) in which they are acting on some body which is not free to accelerate. However, the technique can be of use in measuring some transient forces, and also for calibrating the forces produced by thrust motors in space vehicles.

Force-balance systems

Most force-balance systems are concerned with measuring the force due to gravity on a standard mass and are therefore more correctly described as mass-measuring instruments rather than force-measuring devices. One well known example of this class of device is the beam balance, shown in figure 13.1. Standard masses are added

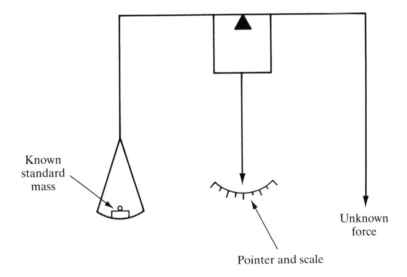

Figure 13.1   The beam balance

to a pan on one side of a pivoted beam until the magnitude of the gravity force on them balances the magnitude of the unknown force applied to the other end of the beam. This equilibrium position is indicated by a pointer which moves against a calibrated scale. Instruments of this type are available for measuring a wide span of masses. Those at the top end of the range would typically measure masses up to 1000 g whereas those at the bottom end of the range can measure masses of less than 0.01 g. Measurement resolution can be as good as 1 part in 10^7 of the full-scale reading if suitable care and expense is taken in the design of the instrument.

Another instrument which works on a force-balance principle is the pendulum scale illustrated in figure 13.2. In this, the unknown force is applied to a platform which is attached by steel tapes to a pair of cams. Downward motion of the platform,

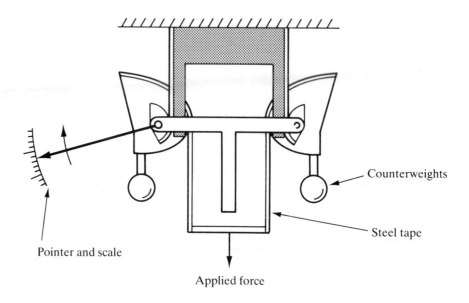

Figure 13.2   The pendulum scale

and hence rotation of the cam, under the influence of the unknown force is opposed by the gravitational force acting on two pendulum-type masses attached to the cams. The amount of rotation of the cams when the equilibrium position is reached is determined by the deflection of a pointer against a scale. This instrument is particularly useful in some applications because it is a relatively simple matter to replace the pointer and scale system by a rotational displacement transducer which gives an electrical output.

Elastic force transducers

The principle of applying a force to an elastic body and measuring the deflection produced is the one most widely used in the task of force measurement. The simplest example of instruments using this physical principle is the spring balance, in which the deflection produced by a force acting on a spring is measured against a calibrated scale. Again, this is essentially a system for measuring mass rather than force.

  The majority of instruments using this principle contain an elastic element of a special shape and design. Some examples of these elastic elements are shown in figure 13.3. Particular aims in the design of these are to obtain a linear output relationship between applied force and measured deflection and to make the instrument insensitive to forces which are not applied directly along the sensing axis. Various types of displacement transducer are used to measure the deflection of elastic elements, selected mainly according to the range of displacements produced in each particular instrument.

  One problem which can affect the performance of this class of instrument is the phenomenon of creep. Creep describes the permanent deformation which an elastic element undergoes after it has been under load for a period of time. This can lead to

Proof ring

Proving frame

Dumbbell-cut proving frame

Solid
cylinder

Rectangular

Figure 13.3 Elastic elements

significant measurement errors in the form of a bias on all readings if the instrument is not recalibrated from time to time. However, careful design and choice of materials can largely eliminate the problem.

The best accuracy in elastic force transducers is obtained by those instruments which use strain gauges to measure displacements of the elastic element, with a figure of ±0.05% of full-scale reading being obtainable. Instruments of this type are used to measure forces over a very wide range between 0 and 3×10^6 N. The typical measurement range of an individual instrument would be 0.1–5 N or 10–3000 kN.

Elastic force transducers based on differential transformers (LVDTs) to measure deflections are used to measure forces up to 25 kN. Apart from having a lower maximum force-measuring capability, they are also inferior to strain-gauge-based instruments in terms of their ±0.2% accuracy figure. Their major advantage is their longevity and almost total lack of maintenance requirements.

The final type of displacement transducer used in this class of instrument is the piezoelectric device. Such instruments are used to measure forces in the range $0–10^6$ N. Piezoelectric crystals replace the specially designed elastic member normally used in this class of instrument, allowing the device to be physically small. As discussed previously, such devices can only measure dynamically changing forces because the output reading results from an induced charge whose magnitude leaks away with time. The fact that the elastic element consists of the piezoelectric crystal itself means that it is very difficult to design such instruments to be insensitive to forces applied at an angle to the sensing axis. Therefore, special precautions have to be taken in applying these devices. Although such instruments are relatively cheap, their accuracy is only ±1% of full-scale reading, and they also have a high temperature coefficient.

Conversion of force to pressure (load cells)

Instruments which translate an applied force into a pressure change are commonly known as load cells. They work on either pneumatic or hydraulic principles.

A pneumatic load cell is shown schematically in figure 13.4. Application of a force to the cell causes deflection of a diaphragm acting as a variable restriction in a nozzle flapper mechanism. The output pressure measured in the cell is approximately proportional to the magnitude of the applied force. The instrument requires a flow of air at its input of around 0.25 m³/h at a pressure of 4 bar. Standard cells are available to measure a wide range of forces. For measuring small forces, instruments are available with a full-scale reading of 25 N, whilst at the top of the range, instruments with a full-scale reading of 25 kN are obtainable. Accuracy is typically ±0.5% of full scale in pneumatic load cells.

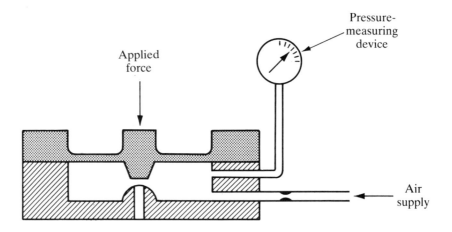

Figure 13.4   The pneumatic load cell

The alternate hydraulic load cell is shown in figure 13.5. In this, the unknown force is applied to oil contained within an enclosed chamber via a diaphragm, and the corresponding increase in oil pressure is measured by a suitable pressure transducer. These instruments are designed to measure much larger forces than pneumatic cells, with a load capacity of 500 kN being common. Special units can be obtained to measure forces as large as 50 000 kN. Besides their much larger measuring range, hydraulic load cells are much more accurate than pneumatic cells, with a figure of ±0.05% of full scale being typical. Measurement resolution is also good at the figure of ±0.02%.

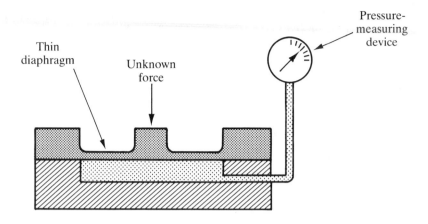

Figure 13.5   The hydraulic load cell

The electromagnetic balance

The electromagnetic balance uses the torque developed by a current-carrying coil suspended in a permanent magnetic field to balance the unknown applied force against the known gravitational force produced on a standard mass, as shown in figure 13.6. A light source and detector system is used to determine the null balance point. The voltage output from the light detector is amplified and applied to the coil,

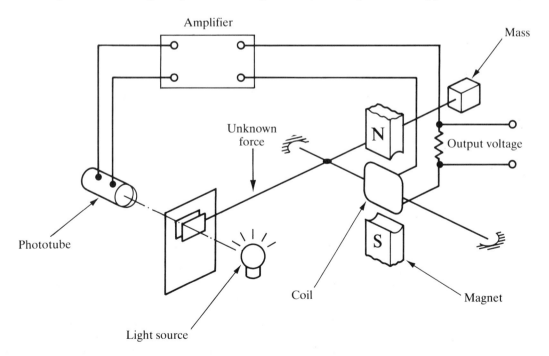

Figure 13.6   The electromagnetic balance

thus creating a servo system where the deflection of the coil in equilibrium is proportional to the applied force. The system is normally used for measuring mass rather than force and so exists as an alternative to the beam balance and the pendulum scale. It does, however, have several advantages over these two mechanical instruments such as its smaller size, its insensitivity to environmental changes (modifying inputs) and its electrical form of output.

Vibrating-wire sensor

This instrument, illustrated in figure 13.7, consists of a wire which is kept vibrating at its resonant frequency by a variable-frequency oscillator. The resonant frequency of a wire under tension is given by:

$$f = \frac{0.5}{L}\left(\frac{M}{T}\right)^{1/2}$$

where M is the mass per unit length of the wire, L is the length of the wire and T is the tension due to the applied force F. Thus, measurement of the output frequency of the oscillator allows the force applied to the wire to be calculated.

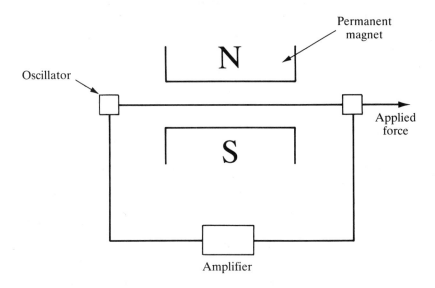

Figure 13.7   The vibrating-wire sensor

Intelligent force transducers

Intelligence within force-measuring systems only exists in the intelligent load cell. The load cell is already a very accurate device and intelligence brings no improvement in this respect. What it does produce is an intelligent weighing system which can compute total cost from the measured weight, using stored cost per unit weight information, and provide an output in the form of a digital display. Cost per weight figures can be prestored for a large number of substances, making the instrument very flexible in its application.

13.2 Torque measurement

Measurement of applied torques is of fundamental importance in all rotating bodies to ensure that the design of the rotating element is adequate to prevent failure under shear stresses. Torque measurement is also a necessary part of measuring the power transmitted by rotating shafts.

The three traditional methods of measuring torque consist of (i) measuring the reaction force in cradled shaft bearings, (ii) the 'Prony brake' method and (iii) measuring the strain produced in a rotating body due to an applied torque. However, recent developments in electronics and optic fiber technology now offer an alternate method, also described below.

Reaction forces in shaft bearings

Any system involving torque transmission through a shaft contains both a power source and a power absorber where the power is dissipated. The magnitude of the transmitted torque can be measured by cradling either the power source or the power absorber end of the shaft in bearings and then measuring the reaction force F and the arm length L, as shown in figure 13.8. The torque is then calculated as the simple product FL. Pendulum scales are very commonly used for measuring the reaction force. Inherent errors in the method are bearing friction and windage torques.

(a) Cradled power source

(b) Cradled power absorber

Figure 13.8   Measuring reaction forces in cradled shaft bearings

Prony brake

The principle of the Prony brake is illustrated in figure 13.9. It is used to measure the

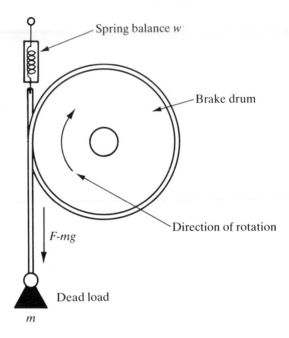

Figure 13.9   The Prony brake

torque in a rotating shaft and consists of a rope wound round the shaft, one end of the rope being attached to a spring balance and the other end carrying a load in the form of a standard mass, m. If the measured force in the spring balance is F_s, then the effective force, F_e, exerted by the rope on the shaft is given by:

$$F_e = mg - F_s$$

If the radius of the shaft is R_s and that of the rope is R_r, then the effective radius, R_e, of the rope and drum with respect to the axis of rotation of the shaft is given by:

$$R_e = R_s + R_r$$

The torque in the shaft, T, can then be calculated as:

$$T = F_e R_e$$

Whilst this is a well known method of measuring shaft torque, a lot of heat is generated because of friction between the rope and shaft, and water cooling is usually necessary.

Measurement of induced strain

Measuring the strain induced in a shaft due to an applied torque has been the most common method used for torque measurement in recent years. It is a very attractive method because it does not disturb the measured system by introducing friction torques in the same way as the last two methods described do. The method involves bonding four strain gauges onto the shaft as shown in figure 13.10, where the strain gauges are arranged in a d.c. bridge circuit. The output from the bridge circuit is a function of the strain in the shaft and hence of the torque applied. It is very important that the positioning of the strain gauges on the shaft is precise, and the difficulty in achieving this makes the instrument relatively expensive.

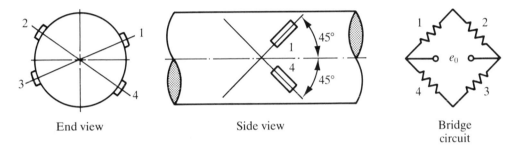

End view   Side view   Bridge circuit

Figure 13.10   Position of torque-measuring strain gauges on shaft

The technique is ideal for measuring the stalled torque in a shaft before rotation commences. However, a problem is encountered in the case of rotating shafts because a suitable method then has to be found for making the electrical connections to the strain gauges. One solution to this problem found in many commercial instruments is to use a system of slip rings and brushes for this, although this increases the cost of the instrument still further.

Optical torque measurement

Optical techniques for torque measurement have become available recently with the development of laser diodes and fiber-optic light transmission systems. One such system is shown in figure 13.11. Two black-and-white striped wheels are mounted at either end of the rotating shaft and are in alignment when no torque is applied to the shaft. Light from a laser diode light source is directed by a pair of optic-fiber cables onto the wheels. The rotation of the wheels causes pulses of reflected light and these are transmitted back to a receiver by a second pair of fiber optic cables. Under zero torque conditions, the two pulse trains of reflected light are in phase with each other. If torque is now applied to the shaft, the reflected light is modulated. Measurement by the receiver of the phase difference between the reflected pulse trains therefore allows the magnitude of torque in the shaft to be calculated. The cost of such instruments is relatively low, and an additional advantage in many applications is their small physical size.

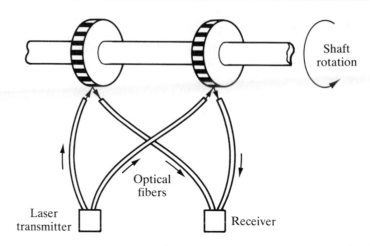

Figure 13.11   Optical torque measurement

14

Pressure Measurement

Pressure measurement is a very common requirement for most industrial process control systems and many different types of pressure-sensing and -measurement systems are available. Before considering these in detail, however, it is important to explain some terms used in pressure measurement and to define clearly the difference between absolute pressure, gauge pressure and differential pressure. The *absolute pressure* of a fluid defines the difference between the pressure of the fluid and the absolute zero of pressure, whereas *gauge pressure* describes the difference between the pressure of a fluid and atmospheric pressure. Absolute and gauge pressures are therefore related by the expression:

Absolute pressure = Gauge pressure + Atmospheric pressure

The term *differential pressure* is used to describe the difference between two absolute pressure values, such as the pressures at two different points within the same fluid (often between the two sides of a flow restrictor in a system measuring volume flow rate).

The range of pressures for which measurement is commonly required is the span from 1.013 to 7000 bar (1 to 6910 atmospheres). In the following discussion, therefore, devices used for measuring pressures in this common mid-range are described first, and this is followed by separate consideration of the techniques used for measuring pressure outside this range.

14.1 Measurement of mid-range pressures

The only instrument commonly used for measuring absolute pressures in the mid-range of 1.013 to 7000 bar is the U-tube manometer, in the form where one end of the U-tube is sealed and evacuated. The U-tube contains a fluid, and the unknown pressure is applied to the open end of the tube (see figure 14.1). The absolute pressure is measured in terms of the difference between the mercury levels in the two halves of the tube. Quite apart from the difficulty in judging exactly where the meniscus levels in the mercury are, such an instrument cannot give a perfect measurement because of the impossibility of achieving a total vacuum at the sealed end of the tube. Although it is possible by modern techniques to design an instrument

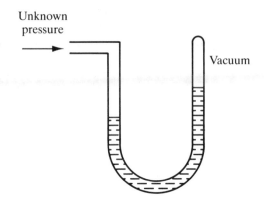

Figure 14.1   Sealed U-tube measuring absolute pressure

which does give a reasonably accurate measurement of absolute pressure, the problem is usually avoided in practice by measuring gauge pressure instead of absolute pressure.

Gauge pressure is generally measured in one of two ways, either by comparison with a known weight acting on a known area, or by deflection of elastic elements. Instruments belonging to the former class are a second form of the U-tube manometer and the dead-weight gauge, whereas the latter class consists of either a diaphragm or some form of Bourdon tube. Apart from these two standard classes of measurement device, modern developments in electronics now allow the use of other principles in pressure measurement, such as in the resonant-wire device.

Choice between the various types of instrument available for measuring mid-range pressures is usually strongly influenced by the intended application. U-tube manometers are very commonly used where visual indication of pressure levels is required, and dead-weight gauges, because of their superior accuracy, are used in calibration procedures of other pressure-measuring devices.

Where compatibility with automatic control schemes is required, the choice of transducer is usually either a diaphragm type or a Bourdon tube. Bellows-type instruments are also used for this purpose but much less frequently, mainly in applications where their greater measurement sensitivity characteristics are required.

U-tube manometer

The U-tube manometer consists of a glass vessel in the shape of a letter U. When used for measuring gauge pressure, both ends of the tube are open, with an unknown pressure being applied at one end and the other end being open to the atmosphere, as shown in figure 14.2(a). The unknown gauge pressure of the fluid (p_1) is then related to the difference (h) in the levels of fluid in the two halves of the tube by the expression:

$$p_1 = h\rho$$

where ρ is the specific gravity of the fluid.

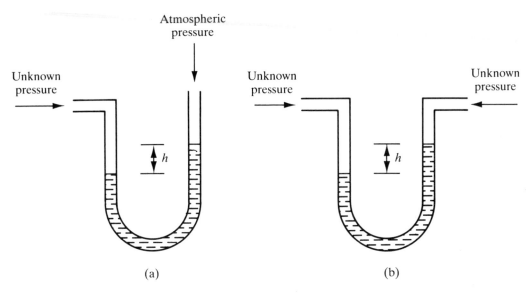

Figure 14.2 (a) U-tube measuring gauge pressure.
(b) U-tube measuring differential pressure

In a third way of using the U-tube manometer, as shown in figure 14.2(b), each open end of the tube is connected to different unknown pressures (p_1 and p_2), and the instrument thereby measures the differential pressure (p_1-p_2) according to the expression:

$$(p_1 - p_2) = h\rho$$

U-tube manometers are typically used to measure gauge and differential pressures up to about 2 bar. The type of liquid used in the instrument depends on the pressure and characteristics of the fluid being measured. Water is a convenient and certainly cheap choice, but evaporates easily and is difficult to see. It is nevertheless used extensively, with the major obstacles to its use being overcome by using colored water and regularly topping up the tube to counteract evaporation.

Situations where water is definitely not used as a U-tube manometer fluid include the measurement of fluids which react with or dissolve in water, and also where higher-magnitude pressure measurements are required. In such circumstances, liquids such as aniline, carbon tetrachloride, bromoform, mercury or transformer oil are used.

The U-tube manometer, in one of its various forms, is an instrument commonly used in industry to give a visual measurement of pressure which can be acted upon by a human operator. It is not normally possible to transform the output of a U-tube manometer into an electrical signal, however, and so this instrument is not suitable for use as part of automatic control systems.

Dead-weight gauge

The dead-weight gauge, as shown in figure 14.3, is a null-reading type of measuring

Figure 14.3   Dead-weight gauge

instrument in which weights are added to the piston platform until the piston is adjacent to a fixed reference mark, at which time the downward force of the weights on top of the piston is balanced by the pressure exerted by the fluid beneath the piston. The fluid pressure is therefore calculated in terms of the weight added to the platform and the known area of the piston. The instrument offers the ability to measure pressures to a high degree of accuracy but is inconvenient to use. Its major application is as a reference instrument against which other pressure-measuring devices are calibrated. Various versions are available which allow measurement of pressures up to 7000 bar.

Diaphragm

The diaphragm is one of the three common types of elastic-element pressure transducer, and is shown schematically in figure 14.4. Diaphragm-type instruments are used to measure pressures up to 10 bar. Applied pressure causes displacement of the diaphragm and this movement is measured by a displacement transducer. Both gauge pressure and differential pressure can be measured by different versions of diaphragm-type instruments. In the case of differential pressure, the two pressures are applied to either side of the diaphragm and the displacement of the diaphragm corresponds to the pressure difference. The typical magnitude of displacement in either version is 0.1 mm, which is well suited to a strain-gauge type of measuring transducer.

Four strain gauges are normally used in a bridge configuration, in which an excitation voltage is applied across two opposite points of the bridge. The output voltage measured across the other two points of the bridge is then a function of the resistance change due to the strain in the diaphragm. This arrangement automatically provides compensation for environmental temperature changes. Older pressure transducers of this type used metallic strain gauges bonded to a diaphragm typically made of stainless steel. Apart from manufacturing difficulties arising from the problem of bonding the gauges, metallic strain gauges have a low gauge factor,

Translational movement

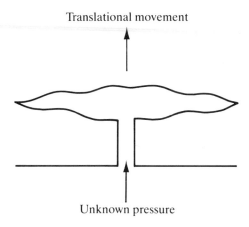

Unknown pressure

Figure 14.4   Diaphragm

which means that the low output from the strain-gauge bridge has to be amplified by an expensive d.c. amplifier. The development of semiconductor (piezoresistive) strain gauges provided a solution to the low-output problem, as they have gauge factors up to 100× greater than metallic gauges. However, the difficulty of bonding gauges to the diaphragm remained and a new problem emerged regarding the highly non-linear characteristic of the strain–output relationship.

The problem of strain-gauge bonding was solved with the emergence of monolithic piezoresistive pressure transducers some 10 years ago, and these are now the most commonly used type of diaphragm pressure transducer. The monolithic cell consists of a diaphragm made of a silicon sheet into which resistors are diffused during the manufacturing process. Besides avoiding the difficulty with bonding, such monolithic silicon measuring cells have the advantage of being very cheap to manufacture in large quantities. Although the inconvenience of a non-linear characteristic remains, this is normally overcome by processing the output signal with an active linearization circuit or incorporating the cell into a microprocessor-based intelligent measuring transducer. The latter usually provide analog-to-digital conversion and interrupt facilities within a single chip and give a digital output which is readily integrated into computer control schemes. Such instruments can also offer automatic temperature compensation, built-in diagnostics and simple calibration procedures. These features allow measurement accuracies of up to ±0.1% of full-scale reading.

As an alternative to strain-gauge-type displacement measurement, capacitive transducers are sometimes used. As a further option, developments in optical fibers have been exploited in the Fotonic sensor (figure 14.5), which is a diaphragm-type device in which the displacement is measured by optoelectronic means. In this device, light travels from a light source down an optical fiber, is reflected back from the diaphragm and travels back along a second fiber to a photodetector. There is a characteristic relationship between the light reflected and the distance from the fiber ends to the diaphragm, thus making the amount of reflected light dependent upon the measured pressure.

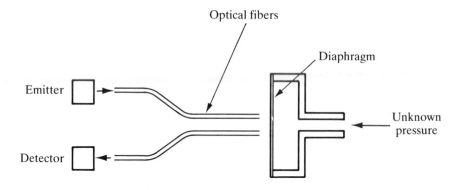

Figure 14.5   Fotonic sensor

Bellows

The bellows, illustrated in figure 14.6, is a device which operates on a very similar principle to the diaphragm, although its use is much less common. The sensitivity of a bellows is greater than for a diaphragm, which is its principal attribute. Pressure changes within the bellows produce translational motion of the end of the bellows which can be measured by capacitive, inductive (LVDT) or potentiometric transducers according to the range of movement produced. A typical measuring range for a bellows-type instrument is 0–1 bar (gauge pressure).

Figure 14.6   Bellows

Bourdon tube

The Bourdon tube is the third type of elastic-element pressure transducer and is a very common industrial measuring instrument used for measuring the pressure of both gaseous and liquid fluids. It consists of a specially shaped piece of oval-section flexible tube which is fixed at one end and free to move at the other. When pressure is applied at the open, fixed end of the tube, the oval cross-section becomes more circular. As the cross-section of the tube tends toward a circular shape, a deflection of the closed, free end of the tube is caused. This displacement is measured by some form of displacement transducer, which is commonly a potentiometer or LVDT, or less often a capacitive sensor. In yet another version, the displacement is measured optically.

The three common shapes of Bourdon tube are shown in figure 14.7. The maximum possible deflection of the free end of the tube is proportional to the angle subtended by the arc through which the tube is bent. For a C-type tube, the maximum value for this arc is somewhat less than 360°. Where greater measurement sensitivity and resolution are required, spiral and helical tubes are used, where the possible magnitude of the arc subtended is limited only by a practical limit on how many turns it is convenient to have in the helix or spiral. This increased measurement performance is only gained, however, at the expense of a substantial increase in manufacturing difficulty and cost compared with C-type tubes, and is also associated with a large decrease in the maximum pressure which can be measured.

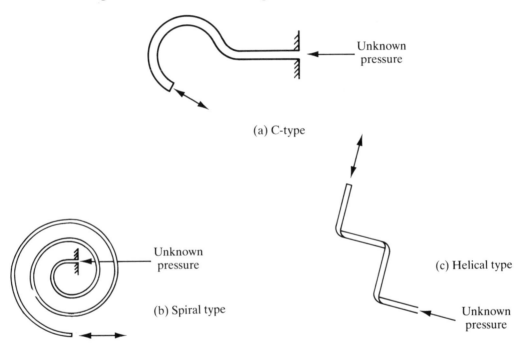

Figure 14.7   Bourdon tubes

C-type tubes are available for measuring pressures up to 6000 bar. A typical C-type tube of 25 mm radius has a maximum displacement travel of 4 mm, giving a moderate level of measurement resolution. Measurement accuracy is typically quoted at ±1% of full-scale deflection. Similar accuracy is available from helical and spiral types, but whilst the measurement resolution is higher, the maximum pressure measurable is only 700 bar.

The existence of one potentially major source of error in Bourdon-tube pressure measurement has not been widely documented and few manufacturers of Bourdon tubes make any attempt to so warn users of their products. The problem is concerned with the relationship between the fluid being measured and the fluid used for calibration. The pointer of Bourdon tubes is normally set at zero during manufacture, using air as the calibration medium. If, however, a different fluid, especially a liquid, is subsequently used with a Bourdon tube, the fluid in the tube will cause a non-zero deflection according to its weight compared with air, resulting in a reading error of up to 6%.

This can be avoided by calibrating the Bourdon tube with the fluid to be measured instead of with air, assuming of course that the user is aware of the problem. Alternately, correction can be made according to the calculated weight of the fluid in the tube. Unfortunately, difficulties arise with both of these solutions if air is trapped in the tube, since this will prevent the tube being filled completely by the fluid. Then, the amount of fluid actually in the tube, and its weight, will be unknown.

In conclusion, therefore, Bourdon tubes only have guaranteed accuracy limits when measuring gaseous pressures. Their use for accurate measurement of liquid pressures poses great difficulty unless the gauge can be totally filled with liquid during both calibration and measurement, a condition which is very difficult to fulfil practically.

Resonant-wire devices

The resonant-wire device is a relatively new instrument which has emanated from recent advances in the electronics field. A typical device is shown schematically in figure 14.8. Wire is stretched across a chamber containing fluid at unknown pressure

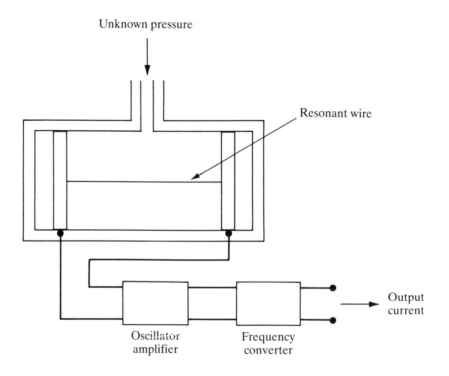

Figure 14.8   Resonant-wire device

subjected to a magnetic field. The wire resonates at its natural frequency according to its tension, which varies with pressure. Thus pressure is calculated by measuring the frequency of vibration of the wire. Such frequency measurement is normally carried out by electronics integrated into the cell. Such devices are highly accurate, with ±0.2% full-scale reading being typical, and they are particularly insensitive to ambient condition changes.

14.2 Low-pressure measurement

Adaptations of most of the common types of pressure transducer already described can be used for absolute pressure measurement in the vacuum range (less than atmospheric pressure). Special forms of Bourdon tubes measure pressures down to 10 mbar, manometers and bellows-type instruments measure pressures down to 0.1 mbar, and diaphragms can be designed to measure pressures down to 0.001 mbar. Other more specialized instruments are also used to measure vacuum pressures. These instruments include the thermocouple gauge, the Pirani gauge, the thermistor gauge, the McLeod gauge and the ionization gauge, and are covered in more detail in the following paragraphs.

As for the case of mid-range pressures, choice between instruments suitable for measuring low pressures depends very much on the application requirements. At the upper end of the low-pressure measurement range (0.001 mbar – 1 bar), Bourdon tubes, manometers, bellows-type and diaphragm instruments are all used depending on whether the application is in process monitoring, automatic control schemes or calibration duties. At greater cost, but generally offering greater sensitivity and accuracy, thermocouple/thermistor, Pirani and McLeod gauges all find application in measuring this range of pressures.

Below pressures of 0.001 mbar, only the more specialized instruments are applicable. Thermocouple/thermistor types measure down to 10^{-4} mbar, Pirani gauges to 10^{-5} mbar and ionization gauges down to 10^{-13} mbar.

Thermocouple gauge

The thermocouple gauge is one of the group of gauges working on the thermal conductivity principle. The Pirani and thermistor gauges also belong to this group. At low pressure, the kinematic theory of gases predicts a linear relationship between pressure and thermal conductivity. Thus measurement of thermal conductivity gives an indication of pressure.

Figure 14.9 shows a sketch of a thermocouple gauge. Operation of the gauge depends on the thermal conduction of heat between a thin hot metal strip in the center and the cold outer surface of a glass tube (which is normally at room tempera-ture). The metal strip is heated by passing a current through it and its temperature is measured by a thermocouple. The temperature measured depends on the thermal conductivity of the gas in the tube and hence on its pressure. A source of error in this instrument is the fact that heat is also transferred by radiation as well as conduction. This is of a constant magnitude, independent of pressure, and so could be measured

Figure 14.9   Thermocouple gauge

and corrected for. However, it is usually more convenient to design for low radiation loss by choosing a heated element with low emissivity. Thermocouple gauges are typically used to measure pressures in the range 10^{-4} mbar up to 1 mbar.

Pirani gauge

A typical form of Pirani gauge is shown in figure 14.10. This is similar to a thermocouple gauge but has a heated element which consists of four coiled tungsten wires

Figure 14.10   Pirani gauge

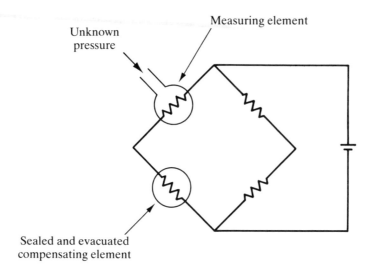

Figure 14.11   Wheatstone bridge circuit to measure output of Pirani gauge

connected in parallel. Two identical tubes are normally used, connected in a bridge circuit as shown in figure 14.11, with one containing the gas at unknown pressure and the other evacuated to a very low pressure. Current is passed through the tungsten element, which attains a certain temperature according to the thermal conductivity of the gas. The resistance of the element changes with temperature and causes an imbalance of the measurement bridge. Thus the Pirani gauge avoids the use of a thermocouple to measure temperature (as in the thermocouple gauge) by effectively using a resistance thermometer as the heated element. Such gauges cover the pressure range 10^{-5} mbar to 1 mbar.

Thermistor gauge

Thermistor gauges operate on identical principles to the Pirani gauge but use semiconductor materials for the heated elements instead of metals. The normal pressure range covered is 10^{-4} mbar to 1 mbar.

McLeod gauge

Figure 14.12 shows the general form of a McLeod gauge, in which low-pressure fluid is compressed to a higher pressure which is then read by manometer techniques. In essence, the gauge can be visualized as a U-tube manometer, sealed at one end, and where the bottom of the U can be blocked at will. To operate the gauge, the piston is first withdrawn, causing the level of mercury in the lower part of the gauge to fall below the level of the junction J between the two tubes in the gauge marked Y and Z. Fluid at unknown pressures p_u is then introduced via the tube marked Z, from where it also flows into the tube marked Y, of cross sectional-area A. Next, the piston is pushed in, moving the mercury level up to block the junction J. At the stage where

Figure 14.12   McLeod gauge

J is just blocked, the fluid in tube Y is at pressure p_u and contained in a known volume V_u. Further movement of the piston compresses the fluid in tube Y and this process continues until the mercury level in tube Z reaches a zero mark. Measurement of the height (h) above the mercury column in tube Y then allows calculation of the compressed volume of the fluid V_c as:

$$V_c = hA$$

Then, by Boyle's law:

$$p_u V_u = p_c V_c$$

Also, applying the normal manometer equation:

$$p_c = p_u + h\rho g$$

where ρ is the mass density of mercury. Hence:

$$p_u = \frac{Ah^2 \rho g}{V_u - Ah} \tag{14.1}$$

The compressed volume V_c is often very much smaller than the original volume, in which case equation (14.1) approximates to:

$$p_u = \frac{Ah^2\rho g}{V_u} \qquad \text{for } Ah \ll V_u \qquad (14.2)$$

Although the best accuracy achievable with McLeod gauges is ±1%, this is still better than most other gauges for measuring pressures in this range, and it is therefore often used as a standard against which other gauges are calibrated. The minimum pressure normally measurable is 10^{-4} mbar, although lower pressures can be measured if pressure-dividing techniques are applied.

Ionization gauge

The ionization gauge is a special type of instrument used for measuring very low pressures in the range 10^{-13} to 10^{-3} bar. Gas of unknown pressure is introduced into a glass vessel containing free electrons discharged from a heated filament, as shown in figure 14.13. Gas pressure is determined by measuring the current flowing between an anode and cathode within the vessel. This current is proportional to the number of ions per unit volume, which in turn is proportional to the gas pressure.

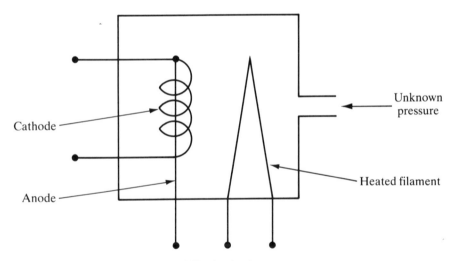

Figure 14.13 Ionization gauge

14.3 High-pressure measurement

Measurement of pressures above 7000 bar is normally carried out electrically by monitoring the change of resistance of wires of special materials. Materials having resistance–pressure characteristics which are suitably linear and sensitive include gold–chromium alloys and manganin. A coil of such wire is enclosed in a sealed, kerosene-filled, flexible bellows, as shown in figure 14.14. The unknown pressure is applied to one end of the bellows, which transmits the pressure to the coil. The magnitude of the applied pressure is then determined by measuring the coil resistance.

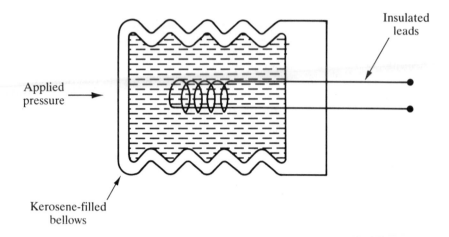

Figure 14.14   High-pressure measurement — wire coil in bellows

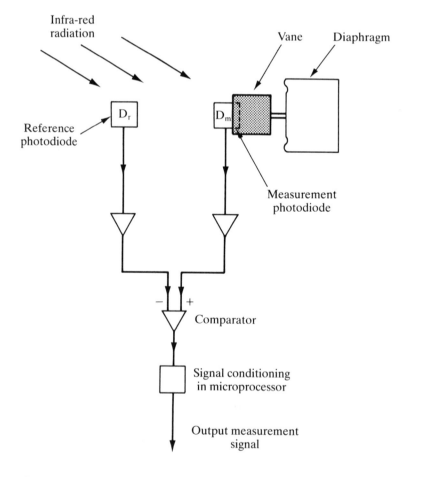

Figure 14.15   One type of intelligent pressure-measuring instrument

14.4 Intelligent pressure transducers

Adding microprocessor power to pressure transducers brings about substantial improvements in their characteristics. Measurement sensitivity improvement, extended measurement range, compensation for hysteresis and other non-linearities, and correction for ambient temperature and pressure changes are just some of the facilities offered by intelligent pressure transducers. Accuracies of $\pm 0.1\%$ can be achieved with piezoresistive-bridge silicon devices, for instance. In view of their much superior characteristics, it is perhaps surprising that intelligent pressure instruments only represent about 1% of the total number of pressure-measuring devices sold at the present time. The significantly higher cost compared with non-intelligent devices has been cited as the reason for this but this hardly seems adequate to explain such a very low level of market penetration.

Some recent microprocessor-based pressure transducers make use of novel techniques of displacement measurement. For example, both diaphragm and helical Bourdon-tube devices are now available which use an optical method of displacement measurement of the form shown in figure 14.15. In this, the motion is transmitted to a vane which progressively shades one of two monolithic photodiodes that are exposed to infra-red radiation. The second photodiode acts as a reference, enabling the microprocessor to compute a ratio signal which is linearized and is available as either an analog or digital measurement of pressure. The typical measurement accuracy is $\pm 0.1\%$.

15

Temperature Measurement

Temperature measurement is very important in all spheres of life and especially so in the process industries. Instruments to measure temperature can be divided into five separate classes according to the physical principle on which they operate. These principles are:

(i)   Thermal expansion.
(ii)   The thermoelectric effect.
(iii)   Resistance change.
(iv)   Resonant frequency change.
(v)   Radiative heat emission.

15.1 Thermal expansion methods

Thermal expansion methods make use of the fact that the dimensions of all substances, whether solids, liquid or gases, change with temperature. Instruments operating on this physical principle include the liquid-in-glass thermometer, the bimetallic thermometer and the pressure thermometer.

Liquid-in-glass thermometers

The liquid-in-glass thermometer is a well known temperature-measuring instrument which is used in a wide range of applications. The fluid used is usually either mercury or colored alcohol, and this is contained within a bulb and capillary tube, as shown in figure 15.1. As the temperature rises, the fluid expands along the capillary tube and the meniscus level is read against a calibrated scale etched on the tube. The process of estimating the position of the curved meniscus of the fluid against the scale introduces some error into the measurement process and a measurement accuracy better than ±1% of full-scale reading is hard to achieve. However, an accuracy of ±0.15% can be obtained in the best industrial instruments. Industrial versions of the liquid-in-glass thermometer are normally used to measure temperature in the range between −200 and +1000°C, although instruments are available to special order which can measure temperatures up to 1500°C.

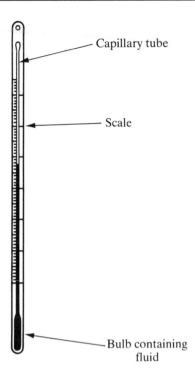

Capillary tube

Scale

Bulb containing
fluid

Figure 15.1   The liquid-in-glass thermometer

Bimetallic thermometer

The bimetallic principle is probably more commonly known in connection with its use in thermostats. It is based on the fact that if two strips of different metals are bonded together, any temperature change will cause the strip to bend, as this is the only way in which the differing rates of change of length of each metal in the bonded strip can be accommodated. In the bimetallic thermostat, this is used as a switch in control applications.

If the magnitude of bending is measured, the bimetallic device becomes a thermometer. For such purposes, the strip is often arranged in a spiral configuration, as shown in figure 15.2, as this gives a relatively large displacement of the free end for any given temperature change. Strips in a helical shape are an alternative for this purpose. The measurement sensitivity is increased further by choosing the pair of materials carefully such that the degree of bending is maximized, with invar (a nickel–steel alloy) and brass being commonly used.

The system used to measure the displacement of the strip must be carefully designed. Very little resistance must be offered to the end of the strip, otherwise the spiral or helix will distort and cause a false reading in the measurement of the displacement. For visual indication purposes, the end of the strip can be made to turn a pointer mounted in low-friction bearings which moves against a calibrated scale. Another suitable form of translational displacement transducer is the LVDT.

Bimetallic thermometers are used to measure temperatures between −75 and

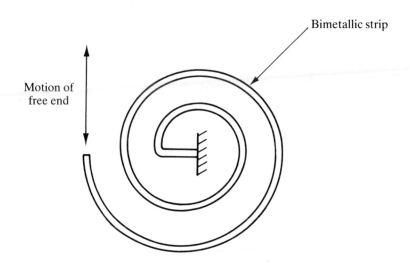

Figure 15.2   The bimetallic thermometer

+1500°C. The accuracy of the best instruments can be as good as +0.5% but such devices are quite expensive. Many instrument applications do not require this degree of accuracy in temperature measurements, and in such cases much cheaper bimetallic thermometers with substantially inferior accuracy specifications are used.

Pressure thermometers

The pressure thermometer measures the variation in pressure of a gas constrained inside a bulb of fixed volume as the temperature changes. As such, it does not strictly belong to the thermal expansion class of instruments but is included because of the relationship between volume and pressure according to Boyle's gas law.

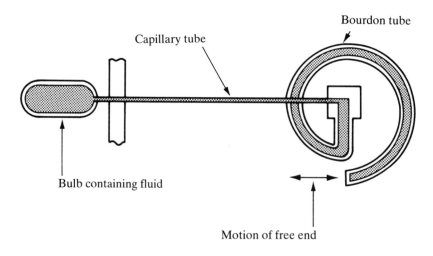

Figure 15.3   The pressure thermometer

The change in pressure of the gas is measured by a suitable pressure transducer such as the Bourdon tube. This transducer is located remotely from the bulb and connected to it by a capillary tube as shown in figure 15.3. The need to protect the pressure-measuring instrument from the environment where the temperature is being measured can require the use of capillary tubes up to 5 m long, and the temperature gradient and hence pressure gradient along the tube acts as a modifying input which can introduce a significant measurement error. Correction for this using the principle of introducing an opposing modifying input, as discussed in chapter 6, can be carried out according to the scheme shown in figure 15.4. This includes a second, dummy capillary tube whose temperature gradient is measured by a second Bourdon tube. The outputs of the two Bourdon tubes are connected together in such a manner that the output from the second tube is subtracted from the output of the first, thus eliminating the error due to the temperature gradient along the tube.

Pressure thermometers are used to measure temperatures in the range between -250 and $+2000°C$. Their typical accuracy is $±0.5\%$ of full-scale reading.

Figure 15.4   Correction for temperature gradient along capillary tube in pressure thermometer

15.2 Thermoelectric-effect instruments (thermocouples)

Thermoelectric-effect instruments rely on the physical principle that, when any two different metals are connected together, an e.m.f., which is a function of the temperature, is generated at the junction between metals. The general form of this relationship is:

$$e = a_1 T + a_2 T^2 + a_3 T^3 + \ldots + a_n T^n$$

This is clearly non-linear, which is inconvenient for measurement applications. Fortunately, for certain pairs of materials, the terms involving squared and higher powers of T (a_2T^2, a_3T^3, etc.) are approximately zero and the e.m.f.–temperature relationship is approximately linear according to:

$$e \simeq a_1 T$$

Wires of such pairs of materials are connected together at one end, and in this form are known as thermocouples. Thermocouples are a very important class of device as they provide the most commonly used method of measuring temperatures in industry.

Thermocouples are manufactured from various combinations of the base metals copper and iron, the base-metal alloys of alumel (Ni/Mn/Al/Si), chromel (Ni/Cr), constantan (Cu/Ni), nicrosil (Ni/Cr/Si) and nisil (Ni/Si/Mn), the noble metals platinum and tungsten, and the noble-metal alloys of platinum/rhodium and tungsten/rhenium. Only certain combinations of these are used as thermocouples and each standard combination is known by an internationally recognized type letter, for instance type K being chromel–alumel. The e.m.f.–temperature characteristics for some of these standard thermocouples are shown in figure 15.5. Certain of these show reasonable linearity over short temperature ranges and their characteristic can therefore be approximated by a series of straight-line relationships for use in

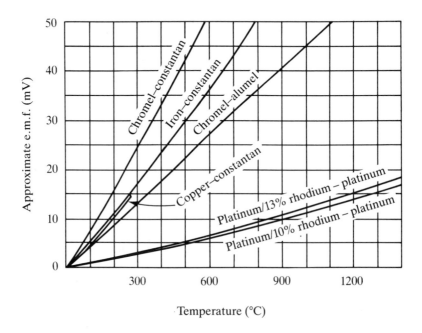

Figure 15.5   E.m.f. – temperature characteristics for some standard thermocouple materials

intelligent instruments containing thermocouples. In general, however, the temperature indicated by a given e.m.f. output measurement has to be calculated from tables. The set of these tables corresponding to all the standard types of thermocouple available are known as *thermocouple tables*.

A typical thermocouple, made from one chromel wire and one constantan wire, is shown in figure 15.6(a). For analysis purposes, it is useful to represent the thermocouple by its equivalent electrical circuit, shown in figure 15.6(b). The e.m.f. generated at the junction is represented by a voltage source, E_1, and it is customary also to show the temperature of the junction on the diagram. This is known as the hot junction and is represented by T_h. The junction between the open ends of the thermocouple and the voltage-measuring instrument is known as the reference junction, and thermocouple tables are calibrated assuming that this is maintained at a temperature of 0°C. This is achieved by immersing the reference junction in an ice bath.

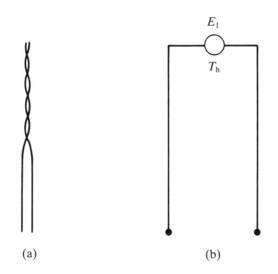

(a)   (b)

Figure 15.6 (a) Typical chromel–alumel thermocouple. (b) Equivalent circuit

Tables for a range of standard thermocouples are given in appendix 4. In these tables, a range of temperatures is given in the left-hand column and the e.m.f. output for each standard type of thermocouple is given in the columns to the right. For instance, if the e.m.f. output measured from a chromel–constantan thermocouple is 13.419 mV with the reference junction at 0°C, the appropriate column in the tables shows that this corresponds to a hot-junction temperature of 200°C. In practice, any general e.m.f. output measurement taken at random would not be found exactly in the tables, and interpolation would be necessary between quoted values. For example, if the measured output e.m.f. for this same type of thermocouple (reference junction at 0°C) was 10.65 mV, it is necessary to carry out linear interpolation between the temperature of 160°C corresponding to an e.m.f. of 10.501 mV shown in the tables and the temperature of 170°C corresponding to an e.m.f. of 11.222 mV. This interpolation procedure gives an indicated hot-junction temperature of 162°C.

Whilst thermocouple tables are calibrated assuming a zero reference-junction temperature, it is not always possible in real applications to maintain such a low reference temperature. In many practical applications of thermocouples, it is necessary to locate the reference junction in a controlled environment maintained by an electrical heating element at a temperature greater than 0°C. Correction for this different reference-junction temperature then has to be made using the *law of intermediate temperatures*. This states that:

$$E_{(T_h, T_0)} = E_{(T_h, T_r)} + E_{(T_r, T_0)} \tag{15.1}$$

where $E_{(T_h, T_0)}$ is the e.m.f. with the junctions at temperatures T_h and T_0, $E_{(T_h, T_r)}$ is the e.m.f with the junctions at temperatures T_h and T_r, $E_{(T_r, T_0)}$ is the e.m.f. with the junctions at temperatures T_r and T_0, T_h is the hot-junction (measured) temperature, T_0 is 0°C and T_r is the non-zero reference-junction temperature generally somewhere between T_0 and T_h.

For example, suppose that the reference junction of a chromel–constantan thermocouple is maintained at a temperature of 80°C and the output e.m.f. measured is 40.102 mV when the hot junction is immersed in a fluid. The quantities given are $T_r = 80°C$ and $E_{(T_h, T_r)} = 40.102\,\text{mV}$. From the tables, $E_{(T_r, T_0)} = 4.983\,\text{mV}$. Now applying equation (15.1), $E_{(T_h, T_0)} = 40.102 + 4.983 = 45.085$ mV. Again referring to the tables, this indicates a fluid temperature of 600°C.

In order to make a thermocouple conform to some precisely defined e.m.f.–temperature characteristic described by standard tables, it is necessary that all metals used are refined to a high degree of pureness and all alloys are manufactured to an exact specification. This makes the materials used expensive, and consequently thermocouples are typically only a few centimeters long. It is clearly impractical to connect a voltage-measuring instrument to measure the thermocouple output in such close proximity to the environment whose temperature is being measured, and therefore extension leads up to several meters long are normally connected between the thermocouple and the measuring instrument. This modifies the equivalent circuit to that shown in figure 15.7. There are now three junctions in the system and consequently three voltage sources, E_1, E_2 and E_3, which add to give a net e.m.f. output E_n given by:

$$E_n = E_1 + E_2 + E_3$$

The e.m.f. of interest for calculating the unknown temperature is E_1. However, only E_n is measured and therefore E_2 and E_3 must be calculated to obtain E_1, which requires that the temperature at the junction between the thermocouples and extension wires be measured. This difficulty can be avoided if the extension wires are manufactured from materials which make the magnitudes of E_2 and E_3 approximately zero. This is most easily achieved by choosing the extension leads of the same basic materials as the thermocouple but manufactured to a lower specification. Such a solution is still prohibitively expensive in the case of noble-metal thermocouples, and it is necessary in this case to search for base-metal extension leads which have a similar thermoelectric behavior to the noble-metal thermocouple. In this form, the

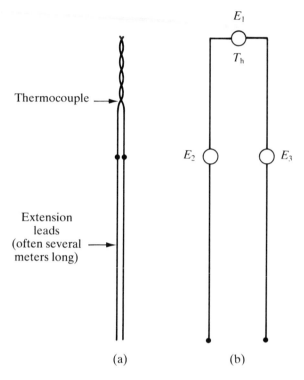

Figure 15.7 (a) Thermocouple with extension leads. (b) Equivalent circuit

extension leads are known as *compensating leads*. A typical example of this is the use of nickel/copper–copper extension leads connected to a platinum/rhodium–platinum thermocouple.

No account so far has been taken of the actual connection between the thermocouple extension leads and the voltage-measuring instrument. This introduces two further junctions between dissimilar metals and hence two extra voltage sources E_4 and E_5, and the total number of voltage sources in the equivalent circuit is increased to five, as shown in figure 15.8. Provided that these extra junctions are maintained at the same temperature, E_4 and E_5 cancel and the net effect on the temperature indicated by the thermocouple tables is zero. For other reference-junction temperatures, correction has to be made for E_4 and E_5, according to the law of intermediate temperatures. Direct application of equation (15.1) is not possible because the junction between a copper connection lead and, say, a chromel extension lead does not involve a standard pair of thermocouple materials and is therefore not covered by the tables. Use of a second thermoelectric behavior law, known as the *law of intermediate metals*, has to be used in these circumstances. This states that the e.m.f. generated at the junction between two metals or alloys A and C is equal to the sum of the e.m.f. generated at the junction between metals or alloys A and B and the e.m.f. generated at the junction between metals or alloys B and C, where all junctions are at the same temperature. This can be expressed more simply as:

$$e_{AC} = e_{AB} + e_{BC} \qquad (15.2)$$

Figure 15.8 (a) Thermocouple with extension leads and connections to a meter.
(b) Equivalent circuit

Suppose that the junction between a copper connection lead and a chromel extension lead is maintained at a temperature of 80°C. Equation (15.2) can be applied by introducing the intermediate alloy, constantan. Then:

$$e_{\text{copper–chromel}} = e_{\text{copper–constantan}} + e_{\text{constantan–chromel}}$$
$$= e_{\text{copper–constantan}} - e_{\text{chromel–constantan}}$$

Therefore, from thermocouple tables:

$$e_{\text{copper–chromel}} = 3.357 - 4.983 = -1.626 \, \text{mV}$$

Thermocouple materials

The five standard base-metal thermocouples are chromel–constantan (type E), iron–constantan (type J), chromel–alumel (type K), nicrosil–nisil (type N) and copper–

constantan (type T). These are all relatively cheap to manufacture but they become inaccurate with age and have a short life.

Chromel–constantan devices give the highest measurement sensitivity of 80 μV/°C, with an accuracy of $\pm0.75\%$ and a useful measuring range of -200 up to 900°C. Unfortunately, whilst they can operate satisfactorily in oxidizing environments, their performance and life are seriously affected by reducing atmospheres. Iron–constantan thermocouples have a sensitivity of 60 μV/°C and are the preferred type for general-purpose measurements in the temperature range of -150 to $+1000$°C, where the typical measurement accuracy obtained is $\pm1\%$. Their performance is little affected by either oxidizing or reducing atmospheres. Copper–constantan devices have a similar measurement sensitivity of 60 μV/°C and find their main application in measuring subzero temperatures down to -200°C, with an accuracy of $\pm0.5\%$. They can also be used in both oxidizing and reducing atmospheres to measure temperatures up to 350°C. Chromel–alumel thermocouples have a measurement sensitivity of only 45 μV/°C, although their characteristic is particularly linear over the temperature range between 700 and 1200°C and this is therefore their main application. Like chromel–constantan devices, they are suitable for oxidizing atmospheres but not for reducing ones. The measurement accuracy given by them is $\pm0.75\%$. Nicrosil–nisil thermocouples are a recent development which resulted from attempts to improve the performance and stability of chromel–alumel thermocouples. Their thermoelectric characteristic has a very similar shape to type-K devices with equally good linearity over the high-temperature measurement range and a measurement sensitivity of 40 μV/°C. The operating environment limitations are the same as for chromel–alumel devices but their long-term stability and life are about three times better.

Noble-metal thermocouples are always expensive but enjoy high stability and long life in conditions of high temperature and oxidizing environments. They are chemically inert except in reducing atmospheres. Thermocouples made from platinum and a platinum/rhodium alloy have a high accuracy of $\pm0.2\%$ and can measure temperatures up to 1500°C, but their measurement sensitivity is only 10 μV/°C. Alternate devices made from tungsten and a tungsten/rhenium alloy have a better sensitivity of 20 μV/°C and can measure temperatures up to 2300°C.

Thermocouple protection

Thermocouples are delicate instruments which must be treated carefully if their specified operating characteristics are to be reproduced. One major source of error is induced strain in the hot junction which reduces the e.m.f. output, and precautions are normally taken to minimize this by mounting the thermocouple horizontally rather than vertically. In some operating environments, no further protection is necessary to obtain satisfactory performance from the instrument. However, thermocouples are prone to contamination by various metals and protection is often necessary to minimize this. Such contamination alters the thermoelectric behavior of the device, such that its characteristic varies from that published in standard tables. It also becomes brittle and its life is therefore shortened.

Protection takes the form of enclosing the thermocouple in a sheath. Some common sheath materials and their maximum operating temperatures are shown in

Table 15.1   Common sheath materials for thermocouples

Material	Maximum operating temperature (°C)†
Mild steel	900
Nickel–chromium	900
Fused silica	1000
Special steel	1100
Mullite	1700
Recrystallized alumina	1850
Beryllia	2300
Magnesia	2400
Zirconia	2400
Thoria	2600

† The maximum operating temperatures quoted assume oxidizing or
neutral atmospheres. For operation in reducing atmospheres, the
maximum allowable temperature is usually reduced.

table 15.1. Whilst the thermocouple is an instrument with a naturally first-order type of step response characteristic, the time constant is usually so small as to be negligible when the thermocouple is used unprotected. When enclosed in a sheath, however, the time constant of the combination of thermocouple and sheath is significant. The size of the thermocouple and hence the diameter required for the sheath has a large effect on the importance of this. The time constant of a thermocouple in a 1 mm diameter sheath is only 0.15 s and this has little practical effect in most measurement situations, whereas a larger sheath of 6 mm diameter gives a time constant of 3.9 s, which cannot be ignored so easily.

Thermocouple manufacture

Thermocouples are manufactured by connecting together two wires of different materials, where each material is produced so as to conform precisely with some defined composition specification. This insures that its thermoelectric behavior accurately follows that for which standard thermocouple tables apply. The connection between the two wires is effected by welding, soldering or in some cases just by twisting the wire ends together. Welding is the most common technique used generally, with silver soldering being reserved for copper–constantan instruments.

The diameter of wire used to construct thermocouples is usually in the range between 0.4 and 2 mm. The larger diameters are used where ruggedness and long life are required, although these advantages are gained at the expense of increasing the measurement time constant. In the case of noble-metal thermocouples, the use of large-diameter wire incurs a substantial cost penalty. Some special applications have a requirement for a very fast response time in the measurement of temperature, and in such cases wire diameters as small as 0.1 μm can be used.

The thermopile

The thermopile is the name given to a temperature-measuring device which consists of several thermocouples connected together in series, such that all the reference junctions are at the same cold temperature and all the hot junctions are exposed to the temperature being measured, as shown in figure 15.9. The effect of connecting n thermocouples together in series is to increase the measurement sensitivity by a factor of n. A typical thermopile manufactured by connecting together 25 chromel–constantan thermocouples gives a measurement resolution of 0.001°C.

Figure 15.9   The thermopile

15.3 Varying-resistance devices

Varying-resistance devices rely on the physical principle of the variation of resistance with temperature. The instruments are known as either resistance thermometers or thermistors according to whether the material used for their construction is a metal or a semiconductor material. The normal method of measuring resistance is to use a d.c. bridge. The excitation voltage of the bridge has to be chosen very carefully because, although a high value is desirable for achieving high measurement sensitivity, the self-heating effect of high currents flowing in the temperature transducer creates an error by increasing the temperature of the device and so changing the resistance value.

Resistance thermometers

Resistance thermometers rely on the principle that the resistance of a metal varies with temperature according to the relationship:

$$R = R_0(1 + a_1 T + a_2 T^2 + a_3 T^3 + \ldots + a_n T^n) \tag{15.3}$$

This equation is non-linear and so is inconvenient for measurement purposes. The equation becomes linear if all the terms in $a_2 T^2$ and higher powers of T are negligible. This is approximately true for some metals over a limited temperature range and, in such cases, the resistance and temperature are related according to:

$$R \simeq R_0(1 + a_1 T)$$

Platinum is one such metal where the resistance–temperature relationship is linear within ±0.4% over the temperature range between −200 and +40°C. Even at +1000°C, the quoted accuracy figure is ±1.2%. The characteristics of platinum and other metals are summarized in figure 15.10.

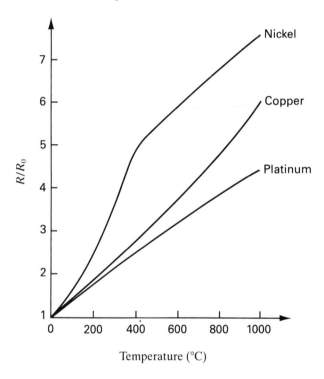

Figure 15.10   Typical resistance – temperature characteristics of metals

Its good linearity and its chemical inertness make platinum the first choice for resistance thermometers in many applications. Platinum is very expensive, however, and consequently the cheaper but less accurate alternatives nickel and copper are sometimes used. These two metals are very susceptible to oxidation and corrosion and so the range of applications where they can be used is strictly limited even if their reduced accuracy is acceptable. Another metal, tungsten, is also used in some circumstances, particularly for high-temperature measurements. The working range of each of these four types of resistance thermometer are as shown below:

Platinum: −270 to 1000°C (though use above 650°C is uncommon)
Copper: −200 to 260°C
Nickel: −200 to 430°C
Tungsten: −270 to 1100°C

In the case of non-corrosive and non-conducting environments, resistance thermometers are used without protection. In all other applications, they are protected inside a sheath. As in the case of thermocouples, such protection affects the speed of response of the system to rapid changes in temperature. A typical time constant for a sheathed platinum resistance thermometer is 0.4 s.

The resistance thermometer exists physically as a coil of resistance wire of the chosen metal. Different instruments available have resistance elements ranging from 10 Ω right up to 25 kΩ. The devices with high resistance have several operational advantages. The first of these is that any connection resistances within the circuit become negligible in their effect. The second advantage is that the relatively high-voltage output produced by these instruments makes any induced e.m.f.'s produced by thermoelectric behavior at the junction with connection leads negligible in magnitude.

Thermistors

Thermistors are manufactured from beads of semiconductor material prepared from oxides of the iron group of metals such as chromium, cobalt, iron, manganese and

Figure 15.11   Typical resistance – temperature characteristics of thermistor materials

nickel. The resistance of such materials varies with temperature according to the following expression:

$$R = R_0 \exp[\beta(1/T - 1/T_0)] \qquad (15.4)$$

This relationship, as illustrated in figure 15.11, exhibits a large negative temperature coefficient (i.e. the resistance decreases as the temperature increases), and so is fundamentally different from the relationship for the resistance thermometer, which shows a positive temperature coefficient. The form of equation (15.4) is such that it is not possible to make a linear approximation to the curve over even a small temperature range, and hence the thermistor is very definitely a non-linear instrument.

The major advantages of thermistors are their relatively low cost and their small size. This size advantage means that the time constant of thermistors operated in sheaths is small. However, the size reduction also decreases its heat dissipation capability, and so makes the self-heating effect greater. In consequence, thermistors have to be operated at generally lower current levels than resistance thermometers and so the measurement sensitivity afforded is less.

15.4 Quartz thermometers

The quartz thermometer makes use of the principle that the resonant frequency of a material such as quartz is a function of temperature, and thus enables temperature changes to be translated into frequency changes. The temperature-sensing element consists of a quartz crystal enclosed within a probe (sheath). The crystal is connected electrically so as to form the resonant element within an electronic oscillator. Measurement of the oscillator frequency therefore allows the measured temperature to be calculated.

The instrument has a very linear output characteristic over the temperature range between -40 and $+230°C$. The measurement resolution given is $0.1°C$.

15.5 Radiation thermometers

All bodies emit electromagnetic radiation as a function of their temperature above absolute zero. The total rate of radiation emission per second is given by:

$$E = KT^4 \qquad (15.5)$$

The power spectral density of this emission varies with temperature in the manner shown in figure 15.12. The major part of the frequency spectrum lies within the band of wavelengths between 0.3 and 40 μm, which corresponds to the visible (0.3–0.72 μm) and infra-red (0.72–1000 μm) ranges. As the magnitude of the radiation varies with temperature, measurement of the emission from a body allows the temperature of the body to be calculated. Choice of the best method of measuring

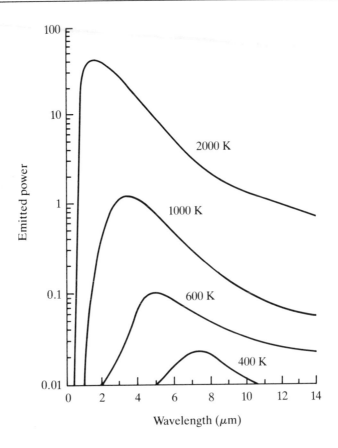

Figure 15.12   Power spectral density of radiated energy emission at
various temperatures

the emitted radiation depends on the temperature of the body. At low temperatures, the peak of the power spectral density function (figure 15.12) lies in the infra-red region, whereas at higher temperatures it moves toward the visible part of the spectrum. This phenomenon is observed as the red glow which a body begins to emit as its temperature is increased beyond 600°C.

Radiation thermometers have one major advantage in that they do not require to be in contact with the hot body in order to measure its temperature. This makes them especially suitable for measuring high temperatures which are beyond the capabilities of contact instruments such as thermocouples, resistance thermometers and thermistors. They are also capable of measuring moving bodies, for instance the temperature of steel bars in a rolling mill. Their use is not as straightforward as the discussion so far might have suggested, however, because the radiation from a body varies with the composition and surface condition of the body as well as with temperature. This dependence on surface condition is quantified by the *emissivity* of the body. The use of radiation thermometers is further complicated by absorption and scattering of the energy between the emitting body and the radiation detector. Energy is scattered by atmospheric dust and water droplets and absorbed by carbon dioxide, ozone and water vapor molecules. Therefore, all radiation thermometers have to be carefully calibrated for each particular body whose temperature they are required to monitor.

Various types of radiation thermometer exist, as described below. The optical pyrometer can only be used to measure high temperatures, but various types of radiation pyrometers are available which between them cover the whole temperature spectrum.

Optical pyrometer

The optical pyrometer, illustrated in figure 15.13, is designed to measure temperatures where the peak radiation emission is in the red part of the visible spectrum, i.e. where the measured body glows a certain shade of red according to the temperature. This limits the instrument to measuring temperatures above 600°C. The instrument contains a heated tungsten filament within its optical system. The current in the filament is increased until its color is the same as the hot body: under these conditions the filament apparently disappears when viewed against the background of the hot body. Temperature measurement is therefore obtained in terms of the current flowing in the filament. Measurement accuracy is improved by employing an optical filter within the instrument which passes a narrow band of frequencies of wavelength around 0.65 μm corresponding to the red part of the visible spectrum.

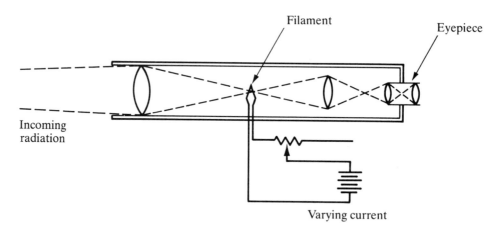

Figure 15.13   The optical pyrometer

The typical measurement accuracy achieved by an optical pyrometer is ±0.5%. The instrument cannot be used in automatic temperature-control schemes because the eye of the human operator is an essential part of the measurement system. It is, nevertheless, widely used in industry for measuring temperatures in furnaces and similar applications.

Radiation pyrometers

All the alternate forms of radiation pyrometer described below have an optical system which is similar to that in the optical pyrometer and focuses the energy emitted from the measured body. They differ, however, by omitting the filament and eyepiece and using instead an energy detector in the same focal plane as the eyepiece was. The radiation detector is either a thermal detector, which measures the temperature rise in a black body at the focal point of the optical system, or a photon detector. Thermal detectors respond equally to all wavelengths in the frequency spectrum whereas photon detectors respond selectively to a particular band within the full spectrum.

Thermopiles, resistance thermometers and thermistors are all used as thermal detectors in different versions of these instruments. These typically have time constants of several milliseconds, because of the time taken for the black body to heat up and the temperature-measuring instrument to respond to the temperature change.

Photon detectors are usually of the photoconductive or photovoltaic type. Both of these respond very much faster to temperature changes than thermal detectors because they involve atomic processes, and typical measurement time constants are a few microseconds.

Various forms of electrical output are available from the radiation detector: these are functions of the incident energy on the detector and are therefore functions of the temperature of the measured body. Whilst this therefore makes such instruments of use in automatic control systems, their accuracy is often inferior to optical pyrometers. This reduced accuracy arises first because a radiation pyrometer is sensitive to a wider band of frequencies than the optical instrument and the relationship between emitted energy and temperature is less well defined. Secondly, the magnitude of energy emission at low temperatures gets very small, according to equation (15.5), increasing the difficulty of accurate measurement.

The forms of radiation pyrometer described below differ mainly in the technique used to measure the emitted radiation. They also differ in the range of energy wavelengths which each is designed to measure and hence in the temperature range measured. One further difference is the material used to construct the energy-focusing lens. Outside the visible part of the spectrum, glass becomes almost opaque to infra-red wavelengths, and other lens materials such as arsenic trisulfide are used.

Broad-band (unchopped) radiation pyrometers

The form of the radiation detector used in this instrument is shown in figure 15.14. The instrument finds wide application in industry and has a measurement accuracy which varies from $\pm 0.05\%$ of full scale in the best instruments to $\pm 0.5\%$ in the cheapest. As its name implies, it measures radiation across the whole frequency spectrum and so uses a thermal detector. This consists of a blackened platinum disk to which a thermopile is bonded. The temperature of the detector increases until the heat gain from the incident radiation is balanced by the heat loss due to convection and radiation. For high-temperature measurement, a two-couple thermopile gives acceptable measurement sensitivity and has a fast time constant of about 0.1 s. At lower measured temperatures, where the level of incident radiation is much less, thermopiles constructed from a greater number of thermocouples must be used to get sufficient measurement sensitivity. This increases the measurement time

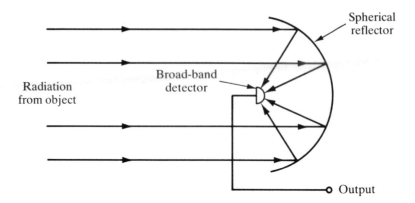

Figure 15.14   Structure of the radiation pyrometer

constant to as much as 2 s. Standard instruments of this type are available to measure temperatures between −20 and +1800°C, although in theory much higher temperatures could be measured by this method.

Chopped broad-band radiation pyrometers

The construction of this form of pyrometer is broadly similar to that shown in figure 15.14 except that a rotary mechanical device is included which periodically interrupts the radiation reaching the detector. The voltage output from the thermal detector thus becomes an alternating quantity which switches between two levels. This form of a.c. output can be amplified much more readily than the d.c. output coming from an unchopped instrument. This is particularly important when amplification is necessary to achieve an acceptable measurement resolution in situations where the level of incident radiation from the measured body is low. For this reason, this form of instrument is the more common when measuring body temperatures associated with peak emission in the infra-red part of the frequency spectrum. For such chopped systems, the time constant of thermopiles is too long. Instead, thermistors are generally used, giving a time constant of 0.01 s. Standard instruments of this type are available to measure temperatures between +20 and +1300°C.

Narrow-band radiation pyrometers

Narrow-band radiation pyrometers use photodetectors of either the photoconductive or photovoltaic form. A photoconductive detector exhibits a change in resistance as the incident radiation level changes whereas a photovoltaic cell exhibits an induced voltage across its terminals which is also a function of the incident radiation level. All photodetectors are preferentially sensitive to a particular narrow band of wavelengths and all have a form of output which varies in a highly non-linear fashion with temperature. Four commonly used materials for photodetectors are cadmium sulfide, lead sulfide, indium antimonide and lead tin telluride. Each of these is sensitive to a different band of wavelengths and therefore all find application in measuring the particular temperature ranges corresponding to each of these bands.

The output from the narrow-band radiation pyrometer is normally chopped into an a.c. signal in the same manner as used in the chopped broad-band pyrometer. This simplifies the amplification of the output signal which is necessary to achieve an acceptable measurement resolution. The typical time constant of a photon detector is only 5 μs which allows high chopping frequencies up to 20 kHz. This gives such instruments an additional advantage in being able to measure fast transients in temperature as short as 10 μs.

Two-color pyrometer

As stated earlier, the emitted radiation–temperature relationship for a body depends on its emissivity. This is very difficult to calculate, and therefore in practice all pyrometers have to be calibrated to the particular body they are measuring. The two-color pyrometer is a system which largely overcomes this problem by using the arrangement shown in figure 15.15. Radiation from the body is split equally into two parts which are applied to separate narrow-band filters. The outputs from the filters consist of radiation within two narrow bands of wavelength λ_1 and λ_2. Detectors sensitive to these frequencies produce output voltages V_1 and V_2 respectively. The ratio of these outputs (V_1/V_2) can be shown (see Bentley 1983) to be a function of temperature and to be independent of the emissivity provided that the two wavelengths λ_1 and λ_2 are close together.

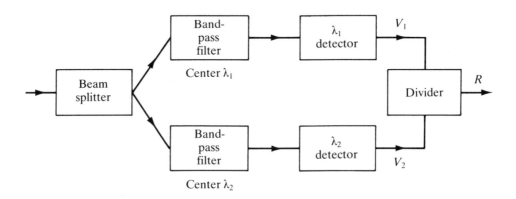

Figure 15.15   Two-color pyrometer system

15.6 Intelligent temperature-measuring instruments

Intelligent temperature transmitters have recently been introduced into the catalogs of some instrument manufacturers, and they bring about the usual benefits associated with intelligent instruments. Such transmitters are separate boxes designed for use with transducers which have either a d.c. voltage output in the milli-volt range or an output in the form of a resistance change. They are therefore suitable for use in conjunction with thermocouples, thermopiles, resistance thermometers, thermistors and broad-band radiation pyrometers. All of the transmitters presently available have non-volatile memories where all constants used in correcting output values for modifying inputs, etc., are stored, thus enabling the instrument to survive power failures without losing such information.

The transmitters now available include adjustable damping, noise rejection, self-adjustment for zero and sensitivity drifts and expanded measurement range. These features allow an accuracy level of ±0.05% of full scale to be specified.

At the present time, intelligent instruments are too new to predict their likely penetration into the market for temperature transducers. Their cost is significantly more than their non-intelligent counterparts, and justification purely on the grounds of their superior accuracy is hard to make. However, their expanded measurement range means immediate savings are made in terms of the reduction in the number of spare instruments needed to cover a number of measurement ranges. Their capability for self-diagnosis and self-adjustment means that they require attention much less frequently, giving additional savings in maintenance costs.

15.7 Choice between temperature transducers

The suitability of different instruments depends substantially on whether the medium to be measured is a solid or a fluid. For measuring the temperature of solids, it is essential that good contact is made between the body and the transducer unless a radiation thermometer is used. This restricts the range of suitable transducers to thermocouples, thermopiles, resistance thermometers and thermistors. On the other hand, fluid temperatures can be measured by any of the instruments described in this chapter, with the exception of radiation thermometers.

The range of instruments working on the thermal expansion principle are mainly used as temperature-indicating devices rather than as components within automatic control schemes. Mercury-in-glass thermometers are used up to +1000°C, bimetallic thermometers to +1500°C and pressure thermometers to +2000°C. The displacement form of output in these latter two devices is frequently connected mechanically to the pointer of a chart recorder in process monitoring applications. The usual measurement accuracy is in the range of ±0.5% to ±1.0%.

The most common instrument used in industry for temperature measurement is the base-metal thermocouple. This is relatively cheap, has a typical accuracy of ±0.5% of full scale and can measure temperatures in the range of −250 to +1200°C. Noble-metal thermocouples are much more expensive, but are chemically inert and can measure temperatures up to 2300°C with an accuracy of ±0.2% of full scale.

Resistance thermometers and thermistors are another relatively common class of devices used for measuring temperature. Resistance thermometers are used in the temperature range of −270 to +1100°C to give a measurement accuracy of ±0.5%. Thermistors are much smaller and cheaper, and give a fast output response to temperature changes, but they have a lower measurement sensitivity than resistance thermometers.

The quartz thermometer is a relatively new development. This is fairly expensive because of the complex electronics required to analyze the frequency-change form of output. It only operates over the limited temperature range of −40 to

+230°C, but gives a high measurement accuracy of ±0.1% within this range.

Radiation thermometers are of considerable benefit in many measurement situations because of their non-invasive mode of measurement. In some applications, the disturbance imposed by putting a transducer into a medium to measure its temperature can be significant. Radiation thermometers are non-contact devices and so avoid such disturbance. Optical pyrometers are commonly used to monitor temperatures above 600°C in industrial furnaces, etc., and give a measurement accuracy of ±0.5%. Various forms of radiation pyrometer are used over the temperature range between −20 and +1800°C and can give measurement accuracies as good as ±0.05%. One particular merit of narrow-band radiation pyrometers is their ability to measure fast temperature transients of duration as small as 10 μs. No other instrument can measure transients anywhere near as fast as this.

The serious difficulty usually encountered in applying radiation thermometers is the need to calibrate the instrument to the body being measured, so correcting for the variable emissivity. This problem is avoided in the two-color pyrometer.

16

Flow Measurement

Flow measurement is extremely important in all the process industries. The manner in which the flow rate is quantified depends on whether the quantity flowing is a solid, liquid or gas. In the case of solids, it is appropriate to measure the rate of mass flow, whereas in the case of liquids and gases, flow is usually measured in terms of the volume flow rate. In a few cases, such as measuring the rate of usage of liquid fuel in a rocket, it is necessary to measure the mass flow rate of a liquid, and special techniques are available for this.

16.1 Mass flow rate

Measurement of the mass flow rate of solids in the process industries is normally

Figure 16.1   Mass-flow-rate measurement

concerned with solids which are in the form of small particles produced by a crushing or grinding process. Such materials are usually transported by some form of conveyor, and this allows the mass flow rate to be calculated in terms of the mass of the material on a given length of conveyor multiplied by the speed of the conveyor.

Figure 16.1 shows a typical system for mass-flow measurement. A load cell measures the mass M of material distributed over a length l of the conveyor. If the conveyor velocity is v, the mass flow rate Q is given by:

$$Q = Mv/l$$

The mass flow rate of a fluid is usually determined by simultaneous measurement of the volume flow rate and the fluid density, although a recently available alternative is the Coriolis meter. As the measurement of the mass flow rate of fluids is not a common requirement, the instruments suitable for this are not covered by this text but full details can be found elsewhere (Medlock 1983, Furness and Heritage 1986).

16.2 Volume flow rate

Volume flow rate is the appropriate way of quantifying the flow of all materials which are in a gaseous, liquid or semi-liquid slurry form (where solid particles are suspended in a liquid host). Materials in these forms are carried in pipes, and the common classes of instrument used for measuring the volume flow rate in such can be summarized as follows:

(i)   Differential pressure meters.
(ii)   Variable-area meters.
(iii)   Positive-displacement meters.
(iv)   Turbine flowmeters.
(v)   Electromagnetic flowmeters.
(vi)   Vortex-shedding flowmeters.
(vii)   Ultrasonic flowmeters.

Consideration of these in the following sections is followed by discussion of the latest developments in cross-correlation, laser Doppler and intelligent flowmeters. The relevant factors governing choice between this bewildering range of available flow-meters in particular applications are considered in the final section of this chapter.

16.3 Differential pressure meters

Differential pressure meters involve the insertion of some device into a fluid-carrying pipe which causes an obstruction and creates a pressure difference on either side of the device. Such devices include the orifice plate, the venturi tube, the flow nozzle, the Dall flow tube and the Pitot tube. When such a restriction is placed in a pipe, the velocity of the fluid through the restriction increases and the pressure decreases. The volume flow rate is then proportional to the square root of the pressure difference across the obstruction. The manner in which this pressure difference is measured is important. Measuring the two pressures with different instruments and calculating the difference between the two measurements is not satisfactory because of the large measurement error which can arise when the pressure difference is small, as explained in chapter 7. The normal procedure is therefore to use a diaphragm-based differential pressure transducer.

All applications of this method of flow measurement assume that flow conditions upstream of the obstruction device are in steady state, and a certain minimum length of straight run of pipe ahead of the flow measurement point is specified to ensure this. The minimum lengths required for various pipe diameters are specified in British Standards tables, but a useful rule of thumb widely used in the process industries is to specify a length of 10 times the pipe diameter. If physical restrictions make this impossible to achieve, special flow-smoothing vanes can be inserted immediately ahead of the measurement point.

Flow-restriction-type instruments are popular because they have no moving parts and are therefore robust, reliable and easy to maintain. One disadvantage of this method is that the obstruction causes a permanent loss of pressure in the flowing fluid. The magnitude and hence importance of this loss depends on the type of obstruction element used, but where the pressure loss is large, it is sometimes necessary to recover the lost pressure by an auxiliary pump further down the flow line. This class of device is not normally suitable for measuring the flow of slurries as the tappings into the pipe to measure the differential pressure are prone to blockage, although the venturi tube can be used to measure the flow of dilute slurries.

Figure 16.2 illustrates approximately the way in which the flow pattern is inter-

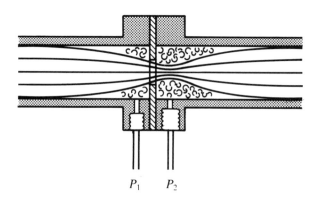

P_1 P_2

Figure 16.2   Profile of flow across orifice plate

rupted when an orifice plate is inserted into a pipe. The other obstruction devices also have a similar effect to this. Of particular interest is the fact that the minimum cross-sectional area of flow occurs not within the obstruction but at a point downstream of there. Knowledge of the pattern of pressure variation along the pipe, as shown in figure 16.3, is also of importance in using this technique of volume-flow-rate measurement. This shows that the point of minimum pressure coincides with the point of minimum cross-section flow, a little way downstream of the obstruction. Figure 16.3 also shows that there is a small rise in pressure immediately before the obstruction. It is therefore important not only to position the instrument measuring P_2 exactly at the point of minimum pressure, but also to measure the pressure P_1 at a point upstream of the point where the pressure starts to rise before the obstruction.

In the absence of any heat transfer mechanisms, and assuming frictionless flow of an incompressible fluid through the pipe, the theoretical volume flow rate of the fluid, Q is given by (see Bentley (1983) for derivation):

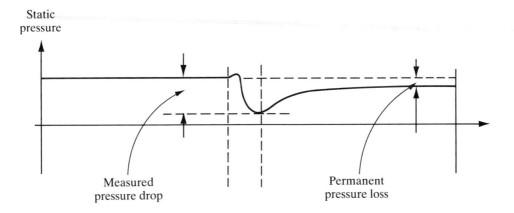

Figure 16.3   Pattern of pressure variation along pipe obstructed by an
orifice plate

$$Q = \frac{A_2}{[1 - (A_2/A_1)^2]^{1/2}}[2(P_1 - P_2)/\rho]^{1/2} \qquad (16.1)$$

where A_1 and P_1 are the cross-sectional area and pressure of the fluid flow before the obstruction, A_2 and P_2 are the cross-sectional area and pressure of the fluid flow at the narrowest point of the flow beyond the obstruction, and ρ is the fluid density.

Equation (16.1) is never applicable in practice for several reasons. First, frictionless flow is never achieved. However, in the case of turbulent flow through smooth pipes, friction is low and it can be adequately accounted for by a variable called the Reynolds number, which is a measurable function of the flow velocity and the viscous friction. The other reasons for the non-applicability of equation (16.1) are that the initial cross-sectional area of the fluid flow is less than the diameter of the pipe carrying it and that the minimum cross-sectional area of the fluid is less than the diameter of the obstruction. Therefore, neither A_1 nor A_2 can be measured. These problems are taken account of by modifying equation (16.1) to the following:

$$Q = \frac{C_D A_2'}{[1 - (A_2'/A_1')^2]^{1/2}}[2(P_1 - P_2)/\rho]^{1/2} \qquad (16.2)$$

where A_1' and A_2' are the pipe diameters before and at the obstruction and C_D is a constant, known as the discharge coefficient, which accounts for the Reynolds number and the difference between the pipe and flow diameters.

Before equation (16.2) can be evaluated, the discharge coefficient must be calculated. As this varies between each measurement situation, it would appear at first sight that the discharge coefficient must be determined by practical experimentation in each case. However, provided that certain conditions are met, standard tables can be used to obtain the value of the discharge coefficient appropriate to the pipe diameter and fluid involved.

It is particularly important in applications of flow restriction methods to choose an instrument whose range is appropriate to the magnitudes of flow rate being measured. This requirement arises because of the square-root type of relationship between the pressure difference and the flow rate, which means that as the pressure difference decreases, the error in flow-rate measurement can become very large. In consequence, restriction-type flowmeters are only suitable for measuring flow rates between 30% and 100% of the instrument range.

Orifice plate

The orifice plate is a metal disk with a hole in it, as shown in figure 16.4, inserted into

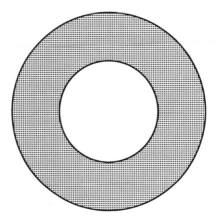

Figure 16.4   The orifice plate

the pipe carrying a flowing fluid. This hole is normally concentric with the disk. Over 50% of the instruments used in industry for measuring volume flow rate are of this type. The use of the orifice plate is so widespread because of its simplicity, cheapness and availability in a wide range of sizes. However, the best accuracy obtainable with this type of obstruction device is only ±2% and the permanent pressure loss caused in the flow is very high, being between 50% and 90% of the pressure difference ($P_1 - P_2$) in magnitude. Other problems with the orifice plate are a gradual change in the discharge coefficient over a period of time as the sharp edges of the hole wear away, and a tendency for any particles in the flowing fluid to stick behind the hole and gradually build up and reduce its diameter. The latter problem can be minimized by using an orifice plate with an eccentric hole. If this hole is close to the bottom of the pipe, solids in the flowing fluid tend to be swept through, and build-up of particles behind the plate is minimal.

A very similar problem arises if there are any bubbles of vapor or gas in the flowing fluid when liquid flow is involved. These also tend to build up behind an orifice plate and distort the pattern of flow. This difficulty can be avoided by mounting the orifice plate in a vertical run of pipe.

Flow nozzle

The form of a flow nozzle is shown in figure 16.5. This is not prone to solid particles

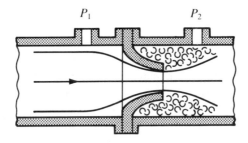

Figure 16.5   The flow nozzle

or bubbles of gas in a flowing fluid sticking in the flow restriction, and so in this respect it is superior to the orifice plate. Its useful working life is also greater because it does not get worn away in the same way as an orifice plate. These factors contribute to giving the instrument a greater measurement accuracy. However, as the engineering effort involved in fabricating a flow nozzle is greater than that required to make an orifice plate, the instrument is somewhat more expensive. In terms of the permanent pressure loss imposed on the measured system, the flow nozzle is very similar to the orifice plate. A typical application of the flow nozzle is in measurement of steam flow.

Venturi

The venturi is a precision-engineered tube of a special shape, as shown in figure 16.6. It is a very expensive instrument but offers very good accuracy and imposes a permanent pressure loss on the measured system of only 10–15% of the pressure difference $(P_1 - P_2)$ across it. The smooth internal shape of this type of restriction means that it is unaffected by solid particles or gaseous bubbles in the flowing fluid, and in fact

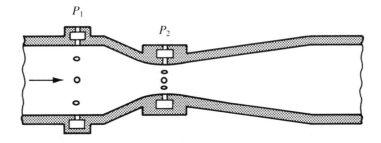

Figure 16.6   The venturi

can even cope with dilute slurries. It has almost no maintenance requirements and its working life is very long.

Dall flow tube

The Dall flow tube, shown in figure 16.7, consists of two conical reducers inserted into the fluid-carrying pipe. It has a very similar internal shape to the venturi, except that it lacks a throat. This construction is much easier to manufacture than a venturi (which requires complex machining), and this gives the Dall flow tube an advantage in cost over the venturi, although the measurement accuracy obtained is not quite as good. Another advantage of the Dall flow tube is its shorter length, which makes the engineering task of inserting it into the flow line easier. The Dall tube has one further operational advantage, in that the permanent pressure loss imposed on the system is only about 5% of the measured pressure difference $(P_1 - P_2)$, and thus is only half that due to a venturi. In other respects, the two instruments are very similar, with their low maintenance requirement and long life.

Figure 16.7   The Dall flow tube

Pitot tube

The Pitot static tube is mainly used for making temporary measurements of flow, although it is also used in some instances for permanent flow monitoring. The instrument depends on the principle that a tube placed with its open end in a stream of fluid, as shown in figure 16.8, will bring to rest that part of the fluid which impinges on it, and the loss of kinetic energy will be converted to a measurable increase in pressure inside the tube.

The flow velocity can be calculated from the formula:

$$v = C[2g(P_1 - P_2)]^{1/2}$$

The constant C, known as the Pitot tube coefficient, is a factor which corrects for the fact that not all fluid incident on the end of the tube will be brought to rest: a proportion will slip around it according to the design of the tube. Having calculated v, the

Figure 16.8   The Pitot tube

volume flow rate can then be calculated by multiplying v by the cross-sectional area of the flow pipe, A.

Inferring the volume flow rate from measurement of the flow velocity at one point in the fluid obviously requires the flow profile to be very uniform. If this condition is not met, multiple Pitot tubes can be used to measure velocities across the whole pipe cross-section.

Pitot tubes have the advantage that they cause negligible pressure loss in the flow. They are also cheap, and the installation procedure consists of the very simple process of pushing them down a small hole drilled in the flow-carrying pipe.

Their main failing is that the measurement accuracy given is only about ±5%, and sensitive pressure-measuring devices are needed to achieve even this limited level of accuracy, as the pressure difference created is very small.

16.4 Variable-area flowmeters

In this class of flowmeter, the differential pressure across a variable aperture is used to adjust the area of the aperture. The aperture area is then a measure of the flow rate. This type of instrument only gives a visual indication of flow rate, and so is of no use in automatic control schemes. However, it is reliable and cheap and used extensively throughout industry. In fact, variable-area meters account for 20% of all flowmeters sold.

In its simplest form, shown in figure 16.9, the instrument consists of a tapered glass tube containing a float which takes up a stable position where its submerged weight is balanced by the upthrust due to the differential pressure across it. The position of the float is a measure of the effective annular area of the flow passage and hence of the flow rate. The accuracy of the cheapest instruments is only ±3%, but more expensive versions offer measurement accuracies as high as ±0.2%. The normal measurement range is between 10% and 100% of the full-scale reading for any particular instrument.

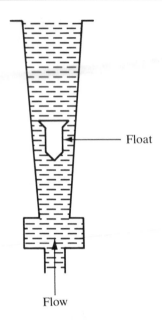

Figure 16.9   The variable-area flowmeter

16.5 Positive-displacement flowmeters

All positive-displacement meters operate by using mechanical divisions to displace discrete volumes of fluid successively. Whilst this principle of operation is common,

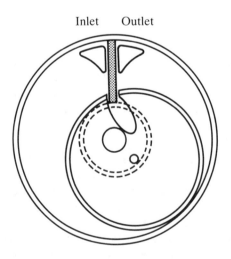

Figure 16.10   Rotary piston form of positive-displacement flowmeter

many different mechanical arrangements exist for putting the principle into practice. All versions of positive-displacement meter are low-friction, low-maintenance and long-life devices, although they do impose a small permanent pressure loss on the flowing fluid. Low friction is especially important when measuring gas flows, and meters with special mechanical arrangements to satisfy this requirement have been developed.

The rotary piston meter is the most common type of positive-displacement meter, and this is illustrated in figure 16.10. This uses a cylindrical piston which is displaced around a cylindrical chamber by the flowing fluid. Rotation of the piston drives an output shaft. This can either be used with a pointer-and-scale system to give a visual flow reading or it can be converted into an electrical output signal.

Positive-displacement flowmeters account for nearly 10% of the total number of flowmeters used in industry. Such devices are used in large numbers for metering domestic gas and water consumption. The cheapest instruments have an accuracy of about ±1.5%, but the accuracy in more expensive ones can be as good as ±0.2%. These higher-quality instruments are used extensively within the oil industry, as such applications can justify the high cost of such instruments.

16.6 Turbine meters

A turbine flowmeter consists of a multi-bladed wheel mounted in a pipe along an axis parallel to the direction of fluid flow in the pipe, as shown in figure 16.11. The flow of fluid past the wheel causes it to rotate at a rate which is proportional to the volume flow rate of the fluid. This rate of rotation is measured by constructing the flowmeter such that it behaves as a variable-reluctance tachogenerator. This is achieved by

Figure 16.11   The turbine flowmeter

fabricating the turbine blades from a ferromagnetic material and placing a permanent magnet and coil inside the meter housing. A voltage pulse is induced in the coil as each blade on the turbine wheel moves past it, and if these pulses are measured by a pulse counter, the pulse frequency and hence flow rate can be deduced. Provided that the turbine wheel is mounted in low-friction bearings, measurement accuracy can be as high as ±0.1%. However, turbine flowmeters are less rugged and reliable than flow-restriction-type instruments, and are badly affected by any particulate matter in the flowing fluid. Bearing wear is a particular problem and they also impose a permanent pressure loss on the measured system.

Turbine meters have a similar cost and market share to positive-displacement meters, and compete for many applications, particularly in the oil industry. Turbine meters are smaller and lighter than the latter and are preferred for low-viscosity, high-flow measurements. Positive-displacement meters are superior, however, in conditions of high viscosity and low flow rate.

16.7 Electromagnetic flowmeters

Electromagnetic flowmeters are limited to measuring the volume flow rate of electrically conductive fluids. A reasonable measurement accuracy, around ±1.5%, is given, although the instrument is expensive both in terms of the initial purchase cost and also in running costs, mainly due to its electricity consumption. A further reason for high cost is the need for careful calibration of each instrument individually during manufacture, as there is considerable variation in the properties of the magnetic materials used.

The instrument, shown in figure 16.12, consists of a stainless-steel cylindrical tube, fitted with an insulating liner, which carries the measured fluid. Typical lining materials used as Neoprene, polytetrafluorethylene (PTFE) and polyurethane. A magnetic field is created in the tube by placing mains-energized field coils either side of it, and the voltage induced in the fluid is measured by two electrodes inserted into opposite sides of the tube. The ends of these electrodes are usually flush with the inner surface of the cylinder. The electrodes are constructed from a material which is unaffected by most types of flowing fluid, such as stainless steel, platinum–iridium alloys, Hastelloy, titanium and tantalum. In the case of the rarer metals in this list, the electrodes account for a significant part of the total instrument cost.

By Faraday's law of electromagnetic induction, the voltage E induced across a length L of the flowing fluid moving at velocity v in a magnetic field of flux density B is given by:

$$E = BLv \qquad (16.3)$$

L is the distance between the electrodes, which is the diameter of the tube, and B is a known constant. Hence, measurement of the voltage E induced across the electrodes allows the flow velocity v to be calculated from equation (16.3). Having thus calculated v, it is a simple matter to multiply v by the cross-sectional area of the tube to obtain a value for the volume flow rate. The typical voltage signal measured across the electrodes is 1 mV when the fluid flow rate is 1 m/s.

Figure 16.12 The electromagnetic flowmeter

The internal diameter of magnetic flowmeters is normally the same as that of the rest of the flow-carrying pipework in the system. Therefore, there is no obstruction to the fluid flow and consequently no pressure loss associated with measurement. Like other forms of flowmeter, the magnetic type requires a minimum length of straight pipework immediately prior to the point of flow measurement in order to guarantee the accuracy of measurement, although a length equal to five pipe diameters is usually sufficient.

Whilst the flowing fluid must be electrically conductive, the method is of use in many applications and is particularly useful for measuring the flow of slurries in which the liquid phase is electrically conductive. At the present time, magnetic flowmeters account for about 15% of the new flowmeters sold and this total is slowly growing. One operational problem is that the insulating lining is subject to damage when abrasive fluids are being handled, and this can give the instrument a limited life.

Current new developments in electromagnetic flowmeters are producing physically smaller instruments and employing better coil designs which reduce electricity consumption. Also, whereas conventional electromagnetic flowmeters require a minimum fluid conductivity of 10 μmho/cm^3, some new versions are now becoming available which can cope with fluid conductivities as low as 1 μmho/cm^3.

16.8 Vortex-shedding flowmeters

Vortex-shedding flowmeters only account for about 1% of flowmeters sold at present, but this percentage is likely to grow in the future as the characteristics of these instruments become more generally known. The operating principle of the instrument is based on the natural phenomenon of vortex shedding, created by placing an unstreamlined obstacle (known as a bluff body) in a fluid-carrying pipe, as indicated in figure 16.13. When fluid flows past the obstacle, boundary layers of viscous, slow-moving fluid are formed along the outer surface. Because the obstacle is not streamlined, the flow cannot follow the contours of the body on the downstream side, and the separate layers become detached and roll into eddies or vortices in the low-pressure region behind the obstacle. The shedding frequency of these alternately shed vortices is proportional to the fluid velocity past the body. Various thermal, magnetic, ultrasonic and capacitive vortex detection techniques are employed in different instruments.

Such instruments have no moving parts, operate over a wide flow range, and require little maintenance. They can measure both liquid and gas flows and a common accuracy figure quoted is ±1% of full-scale reading, though this can be seriously downgraded in the presence of flow disturbances upstream of the measurement point.

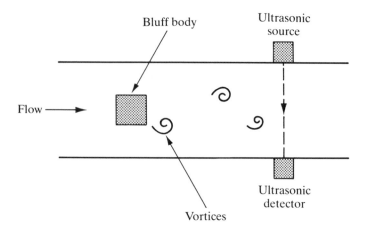

Figure 16.13   The vortex-shedding flowmeter

16.9 Ultrasonic flowmeters

The ultrasonic technique of volume-flow-rate measurement is, like the magnetic flowmeter, a non-invasive method. It is not restricted to conductive fluids, however, and is particularly useful for measuring the flow of corrosive fluids and slurries. A further advantage over magnetic flowmeters is that the instrument is one which clamps on externally to existing pipework rather than being inserted as an integral part of the flow line, as in the case of the magnetic flowmeter. As the procedure of breaking into a pipeline to insert a flowmeter can be as expensive as the cost of the flowmeter itself, the ultrasonic flowmeter has enormous cost advantages. Its clamp-on mode of operation has significant safety advantages in avoiding the possibility of personnel installing flowmeters coming into contact with hazardous fluids such as poisonous, radioactive, flammable or explosive ones. Also, any contamination of the fluid being measured (e.g. food substances and drugs) is avoided. The introduction of this type of flowmeter is a comparatively recent one and their present market share is therefore only about 1% of flowmeters sold. In view of their distinct advantages, however, this proportion is likely to increase over the next few years.

Two different types of ultrasonic flowmeter exist which employ distinct technologies, based on Doppler shift and on transit time. In the past, this has not always been readily understood, and has resulted in ultrasonic technology being rejected entirely when one of these two forms has been found to be unsatisfactory in a particular application. This is unfortunate, because the two technologies have distinct characteristics and areas of application, and many situations exist where one form is very suitable and the other not suitable. To reject both, having only tried out one, is therefore a serious mistake.

Particular care has to be taken to ensure a stable flow profile in ultrasonic flowmeter applications. It is usual to increase the normal specification of the minimum length of straight pipe run prior to the point of measurement, expressed as a number of pipe diameters, from a figure of 10 up to 20 or in some cases even 50 diameters. Analysis of the reasons for poor performance in many instances of ultrasonic flowmeter application has shown failure to meet this stable flow-profile requirement to be a significant factor.

Doppler-shift ultrasonic flowmeter

The principle of operation of the Doppler-shift flowmeter is shown in figure 16.14. A fundamental requirement of these instruments is the presence of scattering elements within the flowing fluid which deflect the ultrasonic energy output from the transmitter such that it enters the receiver. These can be provided by either solid particles, gas bubbles or eddies in the flowing fluid. The scattering elements cause a frequency shift between the transmitted and reflected ultrasonic energy, and measurement of this shift enables the fluid velocity to be inferred.

The instrument consists essentially of an ultrasonic transmitter–receiver pair clamped onto the outside wall of a fluid-carrying vessel. Ultrasonic energy consists of a train of short bursts of sinusoidal waveforms at a frequency between 0.5 and 20 MHz. This frequency range is described as ultrasonic because it is outside the range of human hearing. The flow velocity, v, is given by:

$$v = \frac{c(f_t - f_r)}{2f_t \cos(\theta)} \tag{16.4}$$

where f_t and f_r are the frequencies of the transmitted and received ultrasonic waves respectively, c is the velocity of sound in the fluid being measured, and θ is the angle that the incident and reflected energy waves make with the axis of flow in the pipe. Volume flow rate is then readily calculated by multiplying the measured flow velocity by the cross-sectional area of the fluid-carrying pipe.

The electronics involved in Doppler-shift flowmeters is relatively simple and therefore cheap. Ultrasonic transmitters and receivers are also relatively inexpensive, being based on piezoelectric oscillator technology. As all of its components are

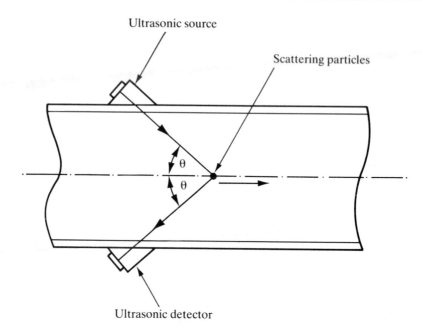

Figure 16.14   Doppler-shift ultrasonic flowmeter

cheap, the Doppler-shift flowmeter itself is inexpensive. The measurement accuracy obtained depends on many factors such as the flow profile, the constancy of pipe-wall thickness, the number, size and spatial distribution of scatterers, and the accuracy with which the speed of sound in the fluid is known. Consequently, accurate measurement can only be achieved by the tedious procedure of carefully calibrating the instrument in each particular flow measurement application. Otherwise measurement errors can approach ±10% of the reading, and for this reason Doppler-shift flowmeters are often used merely as flow indicators, rather than for accurate quantification of the volume flow rate.

Versions are now available which avoid the problem of variable pipe thickness by being fitted inside the flow pipe, flush with its inner surface. Accuracy of ±0.5% is claimed for such devices. Other recent developments are the use of multiple-path ultrasonic flowmeters which use an array of ultrasonic elements to obtain an average velocity measurement which substantially reduces the error due to non-uniform flow profiles. There is a substantial cost penalty involved in this, however.

Transit-time ultrasonic flowmeter

The transit-time ultrasonic flowmeter is an instrument designed for measuring the volume flow rate in clean liquids or gases. It consists of a pair of ultrasonic trans-ducers mounted along an axis aligned at an angle θ with respect to the fluid flow axis, as shown in figure 16.15. Each transducer consists of a transmitter–receiver pair, with the transmitter emitting ultrasonic energy which travels across to the receiver on the opposite side of the pipe. These ultrasonic elements are normally piezoelectric

Figure 16.15   Transit-time ultrasonic flowmeter

oscillators of the same type as used in Doppler-shift flowmeters. Fluid flowing in the pipe causes a time difference between the transit times of the beams traveling upstream and downstream, and measurement of this difference allows the flow velocity to be calculated. The typical magnitude of this time difference is 100 ns in a total transit time of 100 μs, and high-precision electronics are therefore needed to measure it. There are three distinct ways of measuring the time shift. These are direct measurement, conversion to a phase change and conversion to a frequency change. The third of these options is particularly attractive as it obviates the need to measure the speed of sound in the measured fluid which the first two methods require. A scheme applying this third option is shown in figure 16.16. This also multiplexes the transmitting and receiving functions, so that only one ultrasonic element is needed in each transducer.

The forward and backward transit times across the pipe, T_f and T_b, are given by:

$$T_f = \frac{L}{c + v\cos(\theta)} \qquad T_b = \frac{L}{c - v\cos(\theta)}$$

where c is the velocity of sound in the fluid, v is the flow velocity, L is the distance between the ultrasonic transmitter and receiver, and θ is the angle of the ultrasonic beam with respect to the fluid flow axis. The time difference ΔT is given by:

$$\Delta T = T_b - T_f = \frac{2vL\cos(\theta)}{c^2 - v^2\cos^2(\theta)}$$

This requires knowledge of c before it can be solved. However, a solution can be

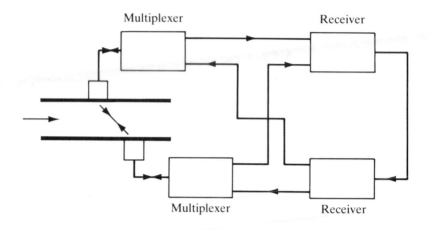

Figure 16.16   Transit-time measurement system

found much more simply if the receipt of a pulse is used to trigger the transmission of the next ultrasonic energy pulse. Then, the frequencies of the forward and backward pulse trains are given by:

$$F_f = \frac{1}{T_f} = \frac{c - v\cos(\theta)}{L} \qquad F_b = \frac{1}{T_b} = \frac{c + v\cos(\theta)}{L}$$

If the two frequency signals are now multiplied together, the resulting beat frequency is given by:

$$\Delta F = F_b - F_f = \frac{v\cos(\theta)}{L}$$

c has now been eliminated and v can be calculated from a measurement of ΔF as:

$$v = \frac{L\,\Delta F}{2\cos(\theta)}$$

Transit-time flowmeters are of more general use than Doppler-shift flowmeters, particularly where the pipe diameter involved is large and hence the transit time is consequently sufficiently large to be measured with reasonable accuracy. It is possible then to achieve an accuracy figure of ±0.5%. The instrument costs more than a Doppler-shift flowmeter, however, because of the greater complexity of the electronics needed to make accurate transit-time measurements.

16.10   Cross-correlation flowmeters

Cross-correlation flowmeters are a type of flowmeter which have not yet achieved widespread practical use in industry. Much development work is still going on, and they therefore mainly exist only as prototypes in research laboratories. However, they are included here because their use is likely to become much more widespread in the future.

Such instruments require some detectable random variable to be present in the flowing fluid. This can take forms such as velocity turbulence and temperature fluctuations. When such a stream of variables is detected by a sensor, the output signal generated consists of noise with a wide frequency spectrum.

Cross-correlation flowmeters use two such sensors placed a known distance apart in the fluid-carrying pipe and cross-correlation techniques are applied to the two output signals from these sensors. This procedure compares one signal with progressively time-shifted versions of the other signal until the best match is obtained between the two waveforms. If the distance between the sensors is divided by this time shift, a measurement of the flow velocity is obtained. A digital processor is an essential requirements to calculate the cross-correlation function, and therefore the instrument must be properly described as an intelligent one.

In practice, the existence of random disturbances in the flow is unreliable, and their detection is difficult. To answer this problem, ultrasonic cross-correlation flowmeters are under development. These use ultrasonic transducers to inject disturbances into the flow and also to detect the disturbances further downstream.

Further information about cross-correlation flowmeters can be found in Medlock (1985).

16.11 Laser Doppler flowmeter

This instrument gives direct measurements of flow velocity for liquids containing suspended particles flowing in a transparent pipe. Light from a laser is focused by an optical system to a point in the flow. The movement of particles causes a Doppler shift of the scattered light and produces a signal in a photodetector which is related to the fluid velocity. A very wide range of flow velocities between 10 μm/s and 800 m/s can be measured by this technique.

Sufficient particles for satisfactory operation are normally present naturally in most liquid and gaseous fluids, and the introduction of artificial particles is rarely needed. The technique is advantageous in measuring flow velocity directly rather than inferring it from a pressure difference. It also causes no interruption in the flow and, as the instrument can be made very small, it can measure velocity in confined areas. One limitation is that it measures local flow velocity in the vicinity of the focal point of the light beam, which can lead to large errors in the estimation of mean volume flow rate if the flow profile is not uniform. However, this limitation is often used constructively in applications of the instrument where the flow profile across the cross-section of a pipe is determined by measuring the velocity at a succession of points.

16.12   Intelligent flowmeters

All the usual benefits associated with intelligent instruments are potentially applicable to many types of flowmeter. However, the availability of such in the market place is currently very limited.

Intelligent differential pressure-measuring instruments can be used to good effect in conjunction with obstruction-type flow transducers. One immediate benefit of this in the case of the commonest flow-restriction device, the orifice plate, is to extend the lowest flow measurable with acceptable accuracy down to 20% of the maximum flow value.

In positive-displacement meters, intelligence allows compensation for thermal expansion of meter components and temperature-induced viscosity changes. Correction for variations in flow pressure is also provided for.

Intelligent electromagnetic flowmeters are also available, and these have a self-diagnosis and self-adjustment capability. The rangeability is 30:1 and quoted accuracy is ±0.5%. It is also normal to include a non-volatile memory to protect constants used for correcting for modifying inputs, etc., against power supply failures.

16.13   Choice between flowmeters for particular applications

The number of relevant factors to be considered when specifying a flowmeter for a particular application is very large. These include the temperature and pressure of the fluid, its density, viscosity, chemical properties and abrasiveness, whether it contains particles, whether it is a liquid or gas, etc. This narrows the field to a subset of instruments which are physically capable of making the measurement. Next, the required performance factors of accuracy, rangeability, acceptable pressure drop, and output signal characteristics must be considered. Finally, the economic viability must be assessed and this must take account not only of purchase cost, but also of reliability, maintenance requirements and service life.

Where only a visual indication of flow rate is needed, the most popular instrument is the variable-area meter. This requirement is common enough for this type of instrument to account for 20% of all flowmeters sold. Where a flow measurement in the form of an electrical signal is required, the choice of available instruments is very large. The orifice plate is used extremely commonly for such purposes and accounts for over 50% of the instruments currently in use in industry. Beyond this very general guidance, the reader is referred elsewhere for more specific information (Furness and Heritage 1986).

17

Level Measurement

A wide variety of instruments are available for measuring the level of liquids and in some cases solids which are in the form of powders or small particles. In most applications of such instruments, only a rough indication of level is needed, and so the accuracy demands are not high.

17.1 Dipsticks

Ordinary dipstick

The ordinary dipstick is perhaps the simplest and cheapest device available for measuring liquid level. It consists of a metal bar, on which a scale is etched as shown in figure 17.1(a), which is fixed at a known position in the liquid-containing vessel. A level measurement is made by removing the instrument from the vessel and reading off how far up the scale the liquid has wetted. As a human operator is required

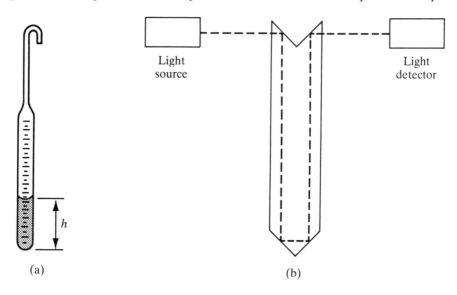

Figure 17.1 (a) Simple dipstick. (b) Optical dipstick

to remove and read the dipstick, this method is limited to use in relatively small and shallow vessels.

Optical dipstick

The optical dipstick, illustrated in figure 17.1(b), allows a reading to be obtained without removing the dipstick from the vessel, and so is applicable to larger, deeper tanks. Light from a source is reflected from a mirror, passes round the chamfered end of the dipstick, and enters a light detector after reflection by a second mirror. When the chamfered end comes into contact with liquid, its internal reflection properties are altered and light no longer enters the detector. By using a suitable mechanical drive system to move the instrument up and down and measure its position, the liquid level can be monitored.

17.2 Float systems

Measuring the position of a float on the surface of a liquid by means of a suitable transducer is another fairly simple and cheap method of liquid level measurement. The system using a potentiometer, shown in figure 17.2, is very common, and is well known for its application to monitoring the level in motor vehicle fuel tanks.

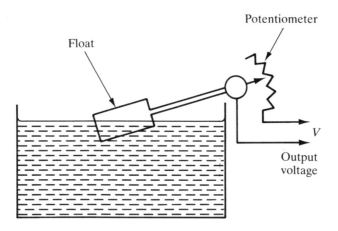

Figure 17.2   Float system

An alternate system called the *float and tape gauge*, which has also been used for a large number of years, it to attach a tape to the float. The tape passes round a pulley situated vertically above the float and its other end is attached to either a counterweight or a negative-rate counterspring. The amount of rotation of the pulley, measured by either a synchro or a potentiometer, is then proportional to the liquid level.

These two essentially mechanical systems of measurement can achieve quite high levels of accuracy, but the maintenance requirements of them are always high.

17.3 Pressure-measuring devices

The hydrostatic pressure due to a liquid is directly proportional to its depth and hence to the level of its surface. Several instruments which use this principle for measuring liquid level are available.

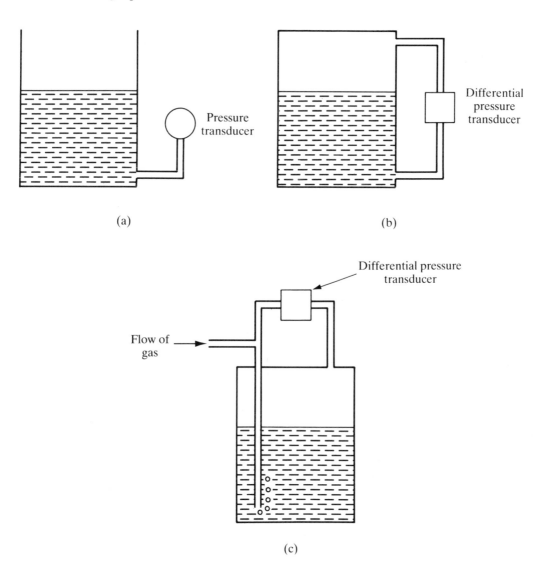

Figure 17.3 (a) Open-topped vessel. (b) Sealed vessel. (c) Bubbler unit

In the case of open-topped vessels (or covered ones which are vented to the atmosphere), the level can be measured by inserting an appropriate pressure transducer at the bottom of the vessel, as shown in figure 17.3(a). The liquid level, h, is then related to the measured pressure, P, according to:

$$h = \frac{P}{\rho g}$$

where ρ is the liquid density and g is the acceleration due to gravity.

One source of error in this method can be imprecise knowledge of the liquid density. This can be a particular problem in the case of liquid solutions and mixtures (especially hydrocarbons), and in some cases only an estimate of density is available. Even with single liquids, the density is subject to variation with temperature, and therefore temperature measurement may be required if very accurate level measurements are needed.

Where liquid-containing vessels are totally sealed, the liquid level can be calculated by measuring the differential pressure between the top and bottom of the tank, as shown in figure 17.3(b). The differential pressure transducer used is normally a diaphragm type, and the liquid level is related to the differential pressure measured, ΔP, according to:

$$h = \frac{\Delta P}{\rho g}$$

The same comments as for the case of the open vessel apply regarding uncertainty in the value of ρ. An additional problem which can occur is an accumulation of liquid on the side of the differential pressure transducer which is measuring the pressure at the top of the vessel. This can arise because of temperature fluctuations, which allow liquid alternately to vaporize from the liquid surface and then condense in the pressure tapping at the top of the vessel. The effect of this on the accuracy of the differential pressure measurement is severe, but the problem is easily avoided by placing a drain pot in the system. This should of course be drained regularly.

A final pressure-related system of level measurement is the bubbler unit shown in figure 17.3(c). This uses a dip pipe which reaches to the bottom of the tank and is purged free of liquid by a steady flow of gas through it. The rate of flow is adjusted until gas bubbles are just seen to emerge from the end of the tube. The pressure in the tube, measured by a pressure transducer, is then equal to the liquid pressure at the bottom of the tank. It is important that the gas used is inert with respect to the liquid in the vessel. Nitrogen or sometimes just air are suitable in most cases. Gas consumption is low, and a cylinder of nitrogen may typically last for six months. The method is suitable for measuring the liquid pressure at the bottom of both open and sealed tanks. It is particularly advantageous in avoiding the large maintenance problem associated with leaks at the bottom of tanks at the site of the pressure tappings required by alternate methods.

17.4 Capacitive devices

Capacitive devices are applicable to measuring the level of both solids in powdered or granular form and liquids. They are suitable for use in extreme applications such as measuring liquid metals (high temperatures), liquid gases (low temperatures),

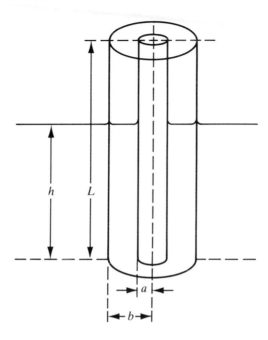

Figure 17.4   Capacitive level sensor

corrosive liquids (acids, etc.) and high-pressure processes. Two versions are used according to whether the measured substance is conducting or not. For non-conducting (<0.1 μmho/cm^3) substances, two bare-metal capacitor plates in the form of concentric cylinders are immersed in the substance, as shown in figure 17.4. The liquid behaves as a dielectric between the plates according to the depth of the liquid. For concentric cylinder plates of radius a and b ($b > a$), and total height L, the depth of the liquid h is related to the measured capacitance C by:

$$h = \frac{C \log_e (b/a) - 2\pi\varepsilon_0}{2\pi\varepsilon_0 (\varepsilon - 1)} \qquad (17.1)$$

where ε is the relative permittivity of the measured substance and ε_0 is the permittivity of free space. The capacitance is measured by one of the methods discussed in the next chapter.

In the case of conducting substances, exactly the same measurement techniques are applied, but the capacitor plates are encapsulated in an insulating material. The relationship between C and h in equation (17.1) then has to be modified to allow for the dielectric effect of the insulator.

17.5 Ultrasonic level gauge

The principle of the ultrasonic level gauge is illustrated in figure 17.5. Energy from an ultrasonic source above the liquid is reflected back from the liquid surface into an

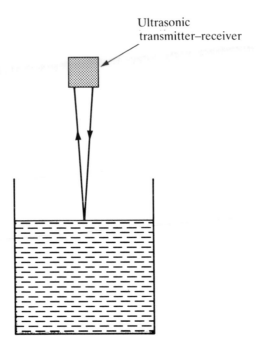

Figure 17.5   Ultrasonic level gauge

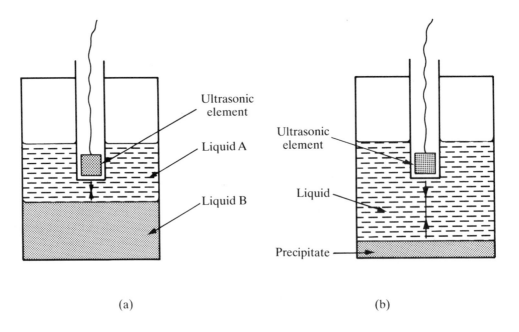

(a)   (b)

Figure 17.6 (a) Measuring liquid/liquid interface.
(b) Measuring liquid/precipitate interface

ultrasonic energy detector. Measurement of the time of flight allows the liquid level to be inferred.

Ultrasonic techniques are especially useful in measuring the position of the interface between two immiscible liquids contained in the same vessel, or measuring the sludge or precipitate level at the bottom of a liquid-filled tank. In either case, the method employed is to fix the ultrasonic transmitter–receiver transducer at a known height in the upper liquid, as shown in figure 17.6. This establishes the level of the liquid/liquid or liquid/sludge level in absolute terms.

17.6 Radiation methods

Radiation methods use a radiation source and detector system located in a liquid-filled tank in the manner shown in figure 17.7. The absorption of both beta-rays and gamma-rays varies with the amount of liquid between the source and detector, and hence is a function of liquid level. Cesium-137 is a gamma-ray source which is commonly used for this purpose. The radiation level measured by the detector, I, is related to the length of liquid in the path, x, according to:

$$I = I_0 \, e^{-\mu \rho x} \tag{17.2}$$

where I_0 is the intensity of radiation which would be received by the detector in the absence of any liquid, μ is the mass absorption coefficient for the liquid and ρ is the mass density of the liquid. In the arrangement shown in figure 17.7, the radiation

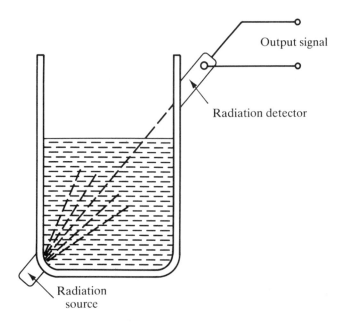

Figure 17.7   Using radiation source to measure level

follows a diagonal path across the liquid, and therefore some trigonometrical manipulation has to be carried out to determine the liquid level, h, from x. In some applications, the ultrasonic source can be located in the center of the bottom of the tank, with the detector vertically above it. Where this is possible, the relationship between the radiation detected and liquid level is obtained by directly substituting h in place of x in equation (17.2). Apart from use with liquids at normal temperatures, this method is commonly used for measuring the level of hot, liquid metals.

17.7 Vibrating level sensor

The principle of the vibrating level sensor is illustrated in figure 17.8. The instrument consists of two piezoelectric oscillators fixed to the inside of a hollow tube which generate flexural vibrations in the tube at its resonant frequency. The resonant frequency of the tube varies according to the depth of its immersion in the liquid. A phase-locked loop circuit is used to track these changes in resonant frequency and adjust the excitation frequency applied to the tube by the piezoelectric oscillators. Liquid level measurement is therefore obtained in terms of the output frequency of the oscillator when the tube is resonating.

Figure 17.8   Vibrating level sensor

17.8 Hot-wire elements

Figure 17.9 shows a level measurement system which uses a series of hot-wire elements placed at regular intervals along a vertical line up the side of a tank. The

Figure 17.9   Hot-wire-element level sensor

heat transfer coefficient of such elements differs substantially depending upon whether the element is immersed in air or in the liquid in the tank. Consequently, elements in the liquid have a different temperature and therefore a different resistance to those in air. This method of level measurement is a simple one, but the measurement resolution is limited to the distance between sensors. Carbon resistors are sometimes used in place of hot-wire elements.

17.9 Intelligent level-measuring instruments

Only certain types of level-measuring device are suitable for use as intelligent instruments. The pressure-measuring devices (section 17.3) are obvious candidates for inclusion within intelligent level-measuring instruments, and consequent improvements in level measurement accuracy are obtained. Such instruments can also carry out additional functions such as providing automatic compensation for liquid density variations. Apart from this, there is little activity at present directed toward developing other forms of intelligent level-measuring systems. This is mainly the result of market forces, as most applications of liquid measurement do not demand high degrees of accuracy but require instead just a rough indication of level.

18

Measurement of Electrical Quantities

18.1 Voltage measurement

Atoms are composed of electrically charged particles, called electrons and protons, which occur in equal numbers in the equilibrium state. If a deficit in either type of particle exists, the atom is said to have either a positive or negative charge, which is measured in units called *coulombs* (C). Such a charge is a source of potential energy and it has the capacity to do work. If such a charge is present at some position within an electrical circuit, it will move to some lower energy level if released, and expend energy in so doing. This net positive or negative charge is called a *potential difference or voltage* and is measured in units called *volts* (V).

   The particular concern in this chapter is not with means of generating voltages, but merely with ways of measuring them. The measuring instruments available include electromechanical (analog) voltmeters, electronic (analog) voltmeters, the cathode ray oscilloscope and digital voltmeters. The first of these, electromechanical voltmeters, are a passive type of instrument, but all the others are active instruments which include signal amplification stages of some form.

   The electromechanical group consists of the moving-iron voltmeter, the moving-coil voltmeter, the electrostatic voltmeter, the dynamometer voltmeter and the thermocouple meter. Most of these have already been discussed in detail in the earlier chapter about techniques for measuring and recording electrical signals in general, and therefore the coverage of them in the following paragraphs is brief. Electronic instruments with an analog form of output are a relatively recent development, and they have some important advantages. First, they have a high input impedance which avoids the circuit loading problems associated with many applications of electromechanical instruments. Secondly, they have an amplification capability which enables them to measure small signal levels accurately. D.c. electronic voltmeters exist as both direct-coupled and chopper-amplifier types. In the case of a.c. versions, three different types exist, known as average-responding, peak-responding and r.m.s.-responding.

   Digital voltmeters have been developed to satisfy a need for higher measurement accuracies and a faster speed of response to voltage changes. As a voltage-indicating instrument, the binary nature of the output reading is readily applied to a display in the form of discrete numerals. Where human operators are required to monitor system voltage levels, this form of output makes an important contribution to measurement reliability and accuracy, since the problem of meter parallax error is eliminated and the possibility of gross error through misreading the meter output is greatly reduced. The availability of a direct output in digital form has also become very valuable to the rapidly expanding range of computer control applications. Digital voltmeters differ mainly in the principle used to effect the analog-to-digital conversion between the measured analog voltage and the output digital reading.

A further class of instruments exists for measuring the very high voltages associated with the electricity supply grid system and certain other systems. The applications for these are so specialized that it is not appropriate to discuss them within this text and the reader is referred elsewhere if such information is required (Buckingham and Price (1966)).

Electromechanical voltmeters

The very earliest instruments used for measuring voltage relied on the principle of interaction between charged bodies. This idea was exploited by Kelvin in the nineteenth century in the development of the electrostatic voltmeter, which measures the attraction forces between two oppositely charged plates. Such forces are relatively small, and consequently this instrument is now only used for measuring d.c. voltages greater than 100 V in magnitude.

The most popular d.c. analog voltmeter used today is the moving-coil type, because of its sensitivity, accuracy and linear scale. This operates at low current levels of 1 mA or so, and hence the basic instrument is only suitable for measuring voltages in the millivolt range and up to around 2 V. For higher voltages, the measuring range of the instrument is increased by placing a resistance in series with the coil, such that a known proportion of the applied voltage is measured by the meter.

Alternate instruments for measuring d.c. voltages are the dynamometer voltmeter and the moving-iron meter. In their basic form, these instruments are most suitable for measuring voltages in the range of 0 – 30 V. Higher voltages are measured by placing a resistance in series with the instrument, as in the case of the moving-coil type.

For measuring a.c. voltages, the usual choice is between the dynamometer voltmeter and the moving-iron meter. Both of these have an r.m.s. form of output. Of the two, the dynamometer voltmeter is more accurate but also more expensive. In both cases, range extension is obtained by using the instruments with a series resistance. This is beneficial to a.c. voltage measurements, as it compensates for the effect of coil inductance by reducing the total resistance/inductance ratio, and hence measurement accuracy is improved. For measurements above about 300 V, it becomes impractical to include a suitable resistance within the case of the instrument because of heat dissipation problems, and instead an external resistance is used.

For measurement of high a.c. voltages, an alternative to using one of these instruments with very large resistances is to transform the voltage to a lower value before applying the measuring instrument. This procedure is the recommended industrial practice for measuring voltages above 750 V.

One major limitation in applying analog meters to a.c. voltage measurement is that the maximum frequency measurable directly is low, 2 kHz for the dynamometer

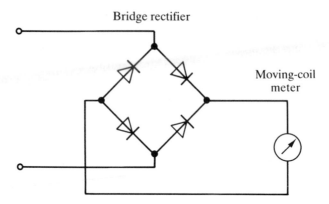

Figure 18.1   Measurement of high-frequency voltage signals

voltmeter and only 100 Hz in the case of the moving-iron instrument. A partial solution to this limitation is to rectify the voltage signal and then apply it to a moving-coil meter. as shown in figure 18.1. This extends the upper measurable frequency limit to 20 kHz. However, the inclusion of the bridge rectifier makes the measurement system temperature-sensitive, and non-linearities significantly affect measurement accuracy for voltages which are small relative to the full-scale value.

An alternate solution to the upper frequency limitation is provided by the *thermocouple meter*, shown in figure 18.2. In this, the a.c. voltage signal heats a small element, and the resulting temperature rise is measured by a thermocouple. The d.c. voltage generated in the thermocouple is applied to a moving-coil meter. The output meter reading is an r.m.s. quantity which varies in a non-linear fashion with the magnitude of the measured voltage. Very high frequencies up to 50 MHz can be measured by this method.

Before ending this section on electromechanical meters, mention must also be made of the *analog multi-meter*, whose working arrangement is shown in figure

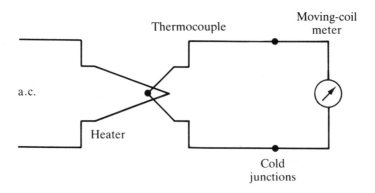

Figure 18.2   The thermocouple meter

Figure 18.3   The analog multi-meter

18.3. This combines a d.c. moving-coil meter with a bridge rectifier to extend its measuring capability to a.c. quantities. A set of rotary switches allows the selection of various series and shunt resistors, which make the instrument capable of measuring both voltage and current over a number of ranges. An internal power source is also often provided to allow it to measure resistances as well. Whilst this instrument is very useful for giving an indication of voltage levels, the compromises in its design which enable it to measure so many different quantities necessarily mean that its accuracy is not as good as instruments which are purpose-designed to measure just one quantity over a single measuring range.

Electronic analog voltmeters

The standard electronic voltmeter for d.c. measurements consists of a simple direct-coupled amplifier and a moving-coil meter, as shown in figure 18.4(a). For measurement of very low-level voltages of a few microvolts, a more sophisticated circuit, known as a chopper amplifier, is used. This is shown in figure 18.4(b). In this, the d.c. input is chopped at a low frequency of around 250 Hz, passed through a blocking capacitor, amplified, passed through another blocking capacitor to remove drift, demodulated, filtered and applied to a moving-coil meter.

Three versions of electronic voltmeter exist for measuring a.c. signals. The *average-responding type* is essentially a direct-coupled d.c. electronic voltmeter with an additional rectifying stage at the input. The output is a measure of the average value of the measured voltage waveform. The second form, known as a *peak-responding type*, has a half-wave rectifier at the input followed by a capacitor. The

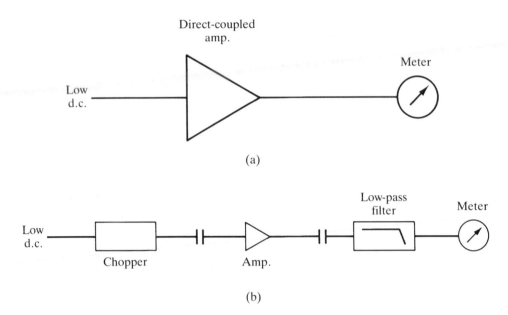

Figure 18.4 (a) Simple d.c. electronic voltmeter.
(b) D.c. electronic voltmeter with chopper amplifier

final part of the circuit consists of an amplifier and moving-coil meter. The capacitor is charged to the peak value of the input signal, and therefore the amplified signal applied to the moving-coil meter gives a reading of the peak voltage in the input waveform. Finally, a third type is available, known as an *r.m.s.-responding type*, which gives an output reading in terms of the r.m.s. value of the input waveform. This type is essentially a thermocouple meter in which an amplification stage has been inserted at the input.

Cathode ray oscilloscope

The principles of operation of the cathode ray oscilloscope have already been discussed in chapter 8. The instrument is widely used for voltage measurement, especially as an item of test equipment for circuit fault finding. It is not a particularly accurate instrument, and is best used where only an approximate measurement is required. Its great value lies in its versatility and applicability to measuring a very wide range of a.c. and d.c. voltages. As a test instrument, it is often required to measure voltages whose frequency and magnitude are totally unknown. The set of rotary switches which alter its range and time base so easily, and the circuitry which protect it from damage when high voltages are applied to it on the wrong range, make it ideally suitable for such applications.

Digital voltmeter (DVM)

The major part of the digital voltmeter is the circuitry that converts the analog voltage being measured into a digital quantity. As the instrument only measures d.c.

quantities in its basic mode, another necessary component within it is one which performs a.c.–d.c. conversion and thereby gives it the capacity to measure a.c. signals. After conversion, the voltage value is displayed by means of indicating tubes or a set of solid-state light-emitting diodes. Four-, five- or even six-figure output displays are commonly used, and although the instrument itself may not be inherently more accurate than some analog types, this form of display enables measurements to be recorded with much greater precision than that obtainable by reading an analog meter scale.

Various techniques are used to effect the analog-to-digital conversion, as discussed in chapter 8. As a general rule, the more expensive and complicated conversion methods achieve a faster conversion speed. The simplest form of DVM is the ramp type. Its main drawbacks are non-linearities in the shape of the ramp waveform used and lack of noise rejection, and these problems limit its accuracy to ±0.05%. It is relatively cheap, however. Another cheap form of DVM is the potentiometric type, and this gives excellent performance for its price. Better noise-rejection capabilities and correspondingly better measurement accuracy (up to ±0.005%) are obtained with integrating-type DVMs. Unfortunately, these are quite expensive.

The digital multi-meter is an extension of the DVM which can measure a.c. and d.c. voltages over a number of ranges through inclusion within it of a set of switchable amplifiers and attenuators. It is widely used in circuit test applications as an alternative to the analog multi-meter, and includes protection circuits which prevent damage if high voltages are applied to the wrong range.

18.2 Current measurement

If atoms with a net positive or negative charge are released from some point in an electrical circuit, the electrically charged particles will flow away from the point of unequal charge distribution until an equilibrium state is reached. The source of charged particles is known as a potential difference, and their rate of flow is known as a current. When current flows in opposite directions along two parallel conductors a force is exerted, and this phenomenon is used to define the unit of *amperes* (A) by which current is measured. When two currents, of magnitude 1 A each, flow in opposite directions along two infinitely long parallel wires positioned 1 m apart in vacuum, the force produced on each conductor is 2×10^{-7} newtons per meter length of conductor.

Any of the current-operated electromechanical voltmeters discussed in section 18.1 can be used to measure current by placing them in series with the current-carrying circuit, and the same frequency limits apply for the measured signal. As in the case of voltage measurement, this upper frequency limit can be raised by rectifying the a.c. current prior to measurement or by using a thermocouple meter. To minimize the effect on the measured system, any current-measuring instrument should have a small resistance. This is opposite to the case of voltage measurement where the instrument is required to have a high resistance for minimal circuit loading.

Moving-coil meters are suitable for direct measurement of d.c. currents in the milliamp range up to 1 A and the range for dynamometer ammeters is up to a few amps. Moving-iron meters can measure d.c. currents up to several hundred amps directly. Similar measurement ranges apply when moving-iron and dynamometer-type instruments are used to measure a.c. currents.

To measure currents higher than allowed by the basic design of electro-mechanical instruments, it is necessary to insert a shunt resistance into the circuit and measure the voltage drop across it. One difficulty which results from this technique is the large power dissipation in the shunt when high-magnitude currents are measured. In the case of a.c. current measurement, care must also be taken to match the resistance and reactance of the shunt to that of the measuring instrument so that frequency and waveform distortion in the measured signal are avoided.

Current transformers provide an alternate method of measuring high-magnitude currents, which avoids the difficulty of designing a suitable shunt. Different versions of these exist for transforming both d.c. and a.c. currents. A d.c. current transformer is shown in figure 18.5. The central d.c. conductor in the instrument is threaded through two magnetic cores which carry two high-impedance windings connected in series opposition. It can be shown (Baldwin 1973) that the current flowing in the windings when excited with an a.c. voltage is proportional to the d.c. current in the central conductor. This output current is commonly rectified and then measured by a moving-coil instrument.

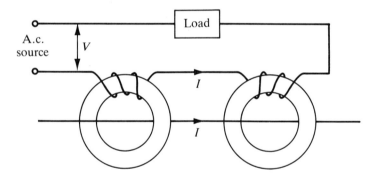

Figure 18.5   D.c. current transformer

An a.c. current transformer typically has a primary winding consisting of only a few copper turns wound on a rectangular or ring-shaped core. The secondary winding on the other hand would normally have several hundred turns according to the current step-down ratio required. The output of the secondary winding is measured by any suitable current-measuring instrument. The design of current transformers is substantially different from that of voltage transformers. The rigidity of its mechanical construction has to be sufficient to withstand the large forces arising from short-circuit currents, and special attention has to be paid to the insulation between its windings for similar reasons. A low-loss core material is used and flux densities are kept as small as possible to reduce losses. In the case of very high currents, the primary winding often consists of a single copper bar which behaves as a single-turn winding. The clamp-on meter, described in chapter 8, is a good example of this.

Apart from electromechanical meters, all the other instruments for measuring voltage discussed in section 18.1 can be applied to current measurement by using them to measure the voltage drop across a known resistance placed in series with the current-carrying circuit. The digital voltmeter and electronic meters are widely applied for measuring currents accurately by this method, and the cathode ray oscilloscope is frequently used to obtain approximate measurements in circuit test applications. Finally, mention must also be made of the use of digital and analog multi-meters for current measurement, particularly in circuit test applications. These instruments include a set of switchable dropping resistors and so can measure currents over a wide range. Protective circuitry within such instruments prevents damage when high currents are applied on the wrong input range.

18.3 Resistance measurement

The standard methods available for measuring resistance, which is measured in units of *ohms* (Ω), include the voltmeter–ammeter method, the resistance-substitution method, the ohmmeter and the d.c. bridge circuit. The digital voltmeter can also be used for measuring resistance if an accurate current source is included within it which passes current through the resistance. This can give a measurement accuracy of $\pm 0.1\%$. Apart from the ohmmeter, these instruments are normally only used to measure medium-value resistances in the range of 1 Ω to 1 MΩ. Special instruments are available for obtaining high-accuracy resistance measurements outside this range (see Baldwin 1973).

Voltmeter–ammeter method

The voltmeter–ammeter method consists of applying a measured d.c. voltage across the unknown resistance and measuring the current flowing. Two alternatives exist for connecting the two meters, as shown in figure 18.6. In figure 18.6(a), the ammeter measures the current flowing in both the voltmeter and the resistance. The error due to this is minimized when the measured resistance is small relative to the voltmeter resistance. In the alternate form of connection, figure 18.6(b), the voltmeter measures the voltage drop across the unknown resistance and the ammeter. Here, the measurement error is minimized when the unknown resistance is large

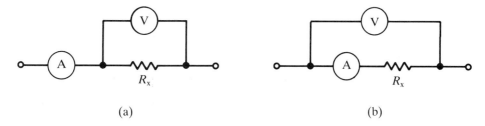

(a)   (b)

Figure 18.6   Voltmeter–ammeter method of measuring resistance with respect to the ammeter resistance. Thus method (a) is best for measurement of small resistances and method (b) for large ones.

Having thus measured the voltage and current, the value of the resistance is then calculated very simply by Ohm's law. This is a suitable method wherever the measurement accuracy of $\pm1\%$ which it gives is adequate.

Resistance-substitution method

In the voltmeter–ammeter method above, either the voltmeter is measuring the voltage across the ammeter as well as across the resistance, or the ammeter is measuring the current flow through the voltmeter as well as through the resistance. The measurement error caused by this is avoided in the resistance-substitution technique. In this method, an unknown resistance in a circuit is temporarily replaced by a variable resistance. The variable resistance is adjusted until the measured circuit voltage and current are the same as existed with the unknown resistance in place. The variable resistance at this point is equal in value to the unknown resistance.

Ohmmeter

The ohmmeter is a simple instrument in which a battery applies a known voltage across a combination of the unknown resistance and a known resistance in series, as shown in figure 18.7. Measurement of the voltage, V_m, across the known resistance, R, allows the unknown resistance, R_u, to be calculated from:

$$R_u = \frac{R(V_b - V_m)}{V_m}$$

where V_b is the battery voltage.

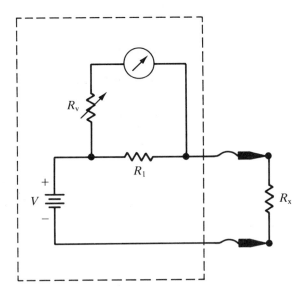

Figure 18.7   The ohmmeter

Ohmmeters are used to measure resistances over a wide range from a few milliohms up to 50 MΩ. The best measurement accuracy is only $\pm2\%$ and they are therefore more suitable for use as test equipment rather than in applications where high accuracy is required. Most of the available versions contain a switchable set of standard resistances, so that measurements of reasonable accuracy over a number of ranges can be made.

Most *digital and analog multi-meters* contain circuitry of the same form as in an ohmmeter, and hence can be similarly used to obtain approximate measurements of resistance.

D.c. bridge circuit

D.c. bridge circuits, as discussed in chapter 4, provide the most commonly used method of measuring medium-value resistances. The best measurement accuracy is provided by the null-output type Wheatstone bridge and figures of $\pm0.02\%$ are achievable with commercially available instruments. Deflection-type bridge circuits are simpler to use in practice than the null-output type, but their measurement accuracy is inferior and the non-linear output relationship is an additional difficulty. One advantage they have is in enabling resistance measurements to be used as inputs to automatic control schemes.

18.4 Power measurement

Measurement of the power consumption in an electrical circuit is needed for many reasons. Circuit components have to be designed to have an adequate power-handling capability, and appropriate means of heat dissipation have to be provided. Such measures are necessary to protect equipment from damage and also to avoid fire hazards due to equipment overheating. Measurement of both the useful power consumption and the power losses in a circuit also allows power utilization efficiency to be optimized.

Power measurement is also important to the electricity supply authorities, to ensure that the power consumption by customers is properly costed and charged for. It is also required to ensure that the total system load is within the capacity of the available generating plant.

Power is measured in units of *watts* (W). The energy required to move 1 C of charge between two energy levels with a potential difference equal to 1 V is called a *joule* (J), and the rate at which this is done is a measure of power expressed in joules per second or in watts.

In the case of d.c. circuits, the power consumption is the simple product of the measured voltage and current. In the case of a.c. circuits, where the voltage and current are varying cyclically but often with a phase difference, it is possible to define a quantity called the instantaneous power which is the product of the voltage level and current level at any instant of time. The instantaneous power varies in a cyclic manner, however, and is of little practical value. Power consumption in a.c. circuits is therefore normally quantified by the average power in watts, which is calculated according to:

$$P_{avg} = V_{rms}I_{rms} \cos(\phi)$$

where ϕ is the phase difference between the current and voltage waveforms.

It is appropriate in some circumstances to measure the reactive power as well, in units of 'volt amps reactive' (V A$_r$). This is given by:

$$P_{reactive} = V_{rms}I_{rms} \sin(\phi)$$

These two quantities are related to the instantaneous power by the vector diagram shown in figure 18.8.

The two common instruments used for measuring power are the dynamometer wattmeter and the electronic wattmeter. Certain other instruments are also used for measurement of high power levels in special applications, such as within the electricity supply industry, as described in Wolf (1973).

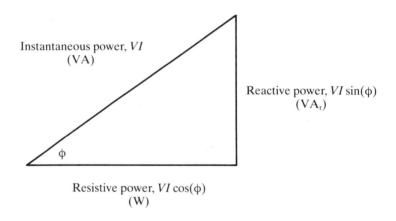

Instantaneous power, *VI*
(VA)

Reactive power, *VI* sin(ϕ)
(VA$_r$)

ϕ

Resistive power, *VI* cos(ϕ)
(W)

Figure 18.8   Vector diagram showing relationship between resistive and reactive power components

Dynamometer wattmeter

The dynamometer wattmeter is the standard instrument used for most power measurements at frequencies below 400 Hz. Products of voltages up to 300 V and currents up to 20 A can be measured. The instrument is the same standard type of dynamometer described in chapter 8 for measuring voltage, except that the coils are connected differently. When used for measuring power dissipation, the stationary coil is connected in series with the measured circuit and carries the circuit current. The rotating coil has a large resistance placed in series with it and is connected across the measured circuit, thus measuring the voltage drop. Theoretically, the instrument measures the product of instantaneous voltage and current and therefore the output pointer reading could be expected to be a measure of instantaneous power, which would make it oscillate at a high frequency. In practice, system inertia prevents such oscillation if the measured circuit has a frequency greater than a few hertz and the pointer settles to a reading proportional to the average circuit power.

Measurement accuracy of commercial instruments varies between ±0.1% and ±0.5% of full scale, according to the design and cost of the instrument. A major source of error in dynamometer wattmeters is power dissipation in the coils. The error due to this becomes very large when the power level in the measured circuit is low. Consequently, the dynamometer wattmeter is not suitable for measuring power levels below 5 W.

Apart from measuring power dissipation, the dynamometer can also measure both the reactive power and the power factor, $\cos(\phi)$, in a circuit. Modifications to the coils and their mode of connection is necessary for this, as described in Wolf (1973).

Electronic wattmeter

The electronic wattmeter is capable of measuring power levels between 0.1 W and 100 kW in circuits operating at frequencies up to 100 kHz. Below 40 kHz, its accuracy is ±3%, but at higher frequencies accuracy can be as poor as ±10%. The instrument is used wherever its accuracy is superior to that of the dynamometer, i.e. at power levels below 5 W and at all frequencies above 400 Hz.

The basic operation of such instruments relies on a circuit containing two amplifiers. One amplifier measures the current through the load and the other measures the voltage across the load. The amplifier outputs are multiplied together by a circuit which uses the logarithmic input relationship of a transistor. The multiplied output is proportional to the instantaneous power level, and is converted to a reading of the average power consumption by the inertial properties of the moving-coil meter to which it is applied. The moving-coil meter ceases to average the instantaneous power properly below a certain frequency, and consequently the electronic wattmeter is not used below frequencies of 10 Hz.

18.5 Frequency measurement

Frequency measurement is required as part of those instruments which convert the measured physical quantity into a frequency change, such as the variable-reluctance velocity transducer, stroboscopes, the vibrating-wire force sensor, resonant-wire pressure-measuring instruments and the quartz thermometer. The output relationship of some forms of a.c. bridge circuit used for measuring inductance and capacitance also requires the excitation frequency to be known.

The instrument which is now most widely used for measuring frequency, which is measured in *hertz* (Hz), is the digital counter–timer. The oscilloscope is also commonly used for obtaining approximate measurements of frequency, especially in circuit test and fault diagnosis applications. Within the audio frequency range, the Wien bridge is a further instrument which is often used. Certain other frequency-measuring instruments also exist which are used in some special applications. These include the dynamometer frequency meter, the zero-beat meter, the grid-dip meter, Campbell's circuit and Maxwell's commutator bridge, and these are discussed in detail elsewhere (Wolf 1973, Baldwin 1973).

Digital counter–timers

Digital counter–timers are the most accurate and flexible instrument available for measuring frequency. An accuracy up to 1 part in 10^8 can be obtained, and all frequencies between d.c. and a few gigahertz can be measured.

The essential component within a counter–timer instrument is an oscillator which provides a very accurately known and stable reference frequency, which is typically either 100 kHz or 1 MHz. This is often maintained in a temperature-regulated environment within the instrument to guarantee its accuracy. The oscillator output is transformed by a pulse-shaper circuit into a train of pulses and applied to an electronic gate, as shown in figure 18.9. Successive pulses at the reference frequency alternately open and close the gate. The input signal of unknown frequency is similarly transformed into a train of pulses and applied to the gate. The number of these pulses which get through the gate during the time that it is open during each gate cycle is proportional to the frequency of the unknown signal.

The accuracy of measurement obviously depends upon how far the unknown frequency is above the reference frequency. As it stands, therefore, the instrument can only accurately measure frequencies which are substantially above 1 MHz. To enable the instrument to measure much lower frequencies, a series of decade frequency dividers are provided within it. These increase the time between the reference-frequency pulses by factors of 10, and a typical instrument can have gate pulses separated in time by between 1 μs and 1 s.

Improvement in the accuracy of low-frequency measurement can be obtained by modifying the gating arrangements such that the signal of unknown frequency is made to control the opening and closing of the gate. The number of pulses at the reference frequency which pass through the gate during the open period is then a measure of the frequency of the unknown signal.

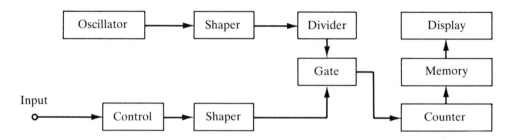

Figure 18.9   Digital counter–timer system

Cathode ray oscilloscope

The cathode ray oscilloscope can be used in two ways to measure frequency. First, the internal time base can be adjusted until the distance between two successive cycles of the measured signal can be read against the calibrated graticule on the screen. Measurement accuracy by this method is limited, but can be optimized by measuring between points in the cycle where the slope of the waveform is steep, generally where it is crossing through from the negative to the positive part of the cycle. Calculation of the unknown frequency from this measured time interval is relatively simple. For example, suppose that the distance between two cycles is 2.5 divisions when the internal time base is set at 10 ms/div. The cycle time is therefore 25 ms and hence the frequency is 1000/25, i.e. 40 Hz. Measurement accuracy is dependent upon how accurately the distance between two cycles is read, and it is very difficult to reduce the error level below ±5% of the reading.

The alternate way of using an oscilloscope to measure frequency is to generate Lissajous patterns. These are produced by applying a known reference-frequency sine wave to the y input (vertical deflection plates) of the oscilloscope and the unknown-frequency sinusoidal signal to the x input (horizontal deflection plates). A pattern is produced on the screen according to the frequency ratio between the two signals, and if the numerator and denominator in the ratio of the two signals both represent an integral number of cycles, the pattern is stationary. Examples of these patterns are shown in figure 18.10, which also shows that phase difference between

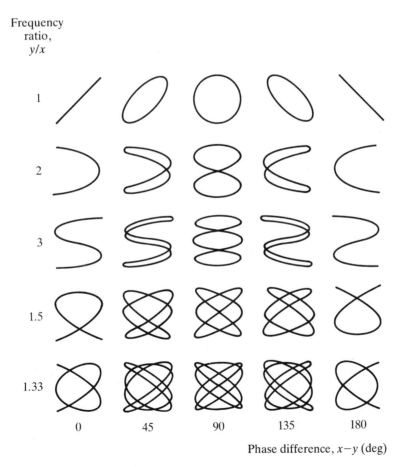

Figure 18.19   Lissajous patterns

the waveforms has an effect on the shape. Frequency measurement proceeds by adjusting the reference frequency until a steady pattern is obtained on the screen and then calculating the unknown frequency according to the frequency ratio which the pattern obtained represents.

Wien bridge

The Wien bridge, shown in figure 18.11, is a special form of a.c. bridge circuit which can be used to measure frequencies in the audio range. An alternate use of the instrument is as a source of audio frequency signals of accurately known frequency.

A simple set of headphones is often used to detect the null-output balance condition. Other suitable instruments for this purpose are the oscilloscope and the electronic voltmeter. At balance, the unknown frequency is calculated according to:

$$f = \frac{1}{2\pi R_3 C_3}$$

The instrument is very accurate at audio frequencies, but at higher frequencies errors due to losses in the capacitors and stray capacitance effects become significant.

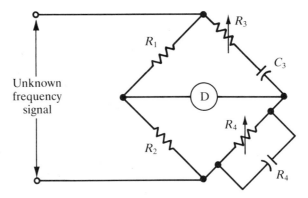

Figure 18.11   The Wien bridge

18.6 Capacitance measurement

The only methods available for accurately measuring capacitance, which is measured in *farads* (F), are the various forms of a.c. bridge described in chapter 4. Various types of capacitance bridge are available commercially, and these would normally be the recommended way of measuring capacitance. In some circumstances, such instruments are not immediately available, and if an approximate measurement of capacitance is acceptable, one of the following two methods can be considered.

The first of these, shown in figure 18.12, consists of connecting the unknown capacitor in series with a known resistance in a circuit excited at a known frequency. An a.c. voltmeter is used to measure the voltage drop across both the resistor and the capacitor. The capacitance value is then given by:

Figure 18.12 Approximate method of measuring capacitance

$$C = \frac{V_r}{2\pi f R V_c}$$

where V_r and V_c are the voltages measured across the resistance and capacitance respectively, f is the excitation frequency and R is the known resistance.

An alternate method of measurement is to measure the time constant of the capacitor connected in an RC circuit.

18.7 Inductance measurement

As in the case of capacitance, inductance, which is measured in *henry* (H), can only be measured accurately by an a.c. bridge circuit. Various commercial inductance bridges are available for this purpose. Such instruments are not always immediately available, however, and the following method can be applied in such circumstances to give an approximate measurement of inductance.

This approximate method consists of connecting the unknown inductance in series with a variable resistance, in a circuit excited with a sinusoidal voltage, as shown in figure 18.13. The variable resistance is adjusted until the voltage measured across the resistance is equal to that measured across the inductance. The two impedances are then equal, and the value of the inductance L can be calculated from:

$$L = \frac{(R^2 - r^2)^{1/2}}{2\pi f}$$

where R is the value of the variable resistance, r is the value of the inductor resistance and f is the excitation frequency.

Figure 18.13   Approximate method of measuring inductance

Conversion Tables Between Imperial and SI Units

Length

SI units: mm, m, km
Imperial units: in, ft, mile

	mm	m	km	in	ft	mile
mm	1	10^{-3}	10^{-6}	0.0393701	3.28×10^{-3}	–
m	1000	1	10^{-3}	39.3701	3.28084	6.21×10^{-4}
km	10^{6}	10^{3}	1	39370.1	3280.84	0.621371
in	25.4	0.0254	–	1	0.083333	–
ft	304.8	0.3048	3.048×10^{-4}	12	1	1.89×10^{-4}
mile	–	1609.34	1.60934	63360	5280	1

Area

SI units: mm^2, m^2, km^2
Imperial units: in^2, ft^2, $mile^2$

	mm^2	m^2	km^2	in^2	ft^2	$mile^2$
mm^2	1	10^{-6}	–	1.55×10^{-3}	1.08×10^{-5}	–
m^2	10^{6}	1	10^{-6}	1550	10.764	–
km^2	–	10^{6}	1	–	1.08×10^{7}	0.3861
in^2	645.16	6.452×10^{-4}	–	1	6.94×10^{-3}	–
ft^2	92903	0.09290	–	144	1	–
$mile^2$	–	2.590×10^{6}	2.590	–	2.788×10^{7}	1

Volume

SI units: mm^3, ml, l, m^3
Imperial units: in^3, ft^3, UK gallon, US gallon

	mm^3	ml	l	m^3	in^3	ft^3	UK gallon	US gallon
mm^3	1	10^{-3}	10^{-6}	10^{-9}	6.10×10^{-5}	–	–	–
ml	10^3	1	10^{-3}	10^{-6}	0.061026	3.53×10^{-5}	2.2×10^{-4}	2.6×10^{-4}
l	10^6	10^3	1	10^{-3}	61.026	0.03532	0.2200	0.2642
m^3	10^9	10^6	10^3	1	61026	35.31	220.0	264.2
in^3	16387	16.39	0.0164	–	1	5.79×10^{-4}	3.61×10^{-3}	4.33×10^{-3}
ft^3	–	2.83×10^{-4}	28.32	0.02832	1728	1	6.229	7.4805
UK gallon	–	4546	4.546	4.55×10^{-3}	277.4	0.1605	1	1.201
US gallon	–	3785	3.785	3.79×10^{-3}	231.0	0.1377	0.8327	1

Force

SI units: N, kg_f, kN
Imperial units: pdl, lb_f, UK ton_f

	N	kg_f	kN	pdl	lb_f	UK ton_f
N	1	0.1020	10^{-3}	7.233	0.2248	1.004×10^{-4}
kg_f	9.807	1	9.81×10^{-3}	70.93	2.2046	9.84×10^{-4}
kN	1000	102.0	1	7233	224.8	0.1004
pdl	0.1383	0.0141	1.38×10^{-4}	1	0.0311	1.39×10^{-5}
lb_f	4.448	0.4536	4.45×10^{-3}	32.174	1	4.46×10^{-4}
UK ton_f	9964	1016	9.964	72070	2240	1

Torque

SI units: N m, kg_f m
Imperial units: pdl ft, lb_f ft

	N m	kg_f m	pdl ft	lb_f ft
N m	1	0.1020	23.73	0.7376
kg_f m	9.807	1	232.7	7.233
pdl ft	0.04214	4.30×10^{-3}	1	0.03108
lb_f ft	1.356	0.1383	32.17	1

Pressure

SI units: mbar, bar, N/m^2
Imperial units: lb/in^2, in Hg, atm

	mbar	bar	N/m^2	lb/in^2	in Hg	atm
mbar	1	10^{-3}	100	0.01450	0.02953	9.869×10^{-4}
bar	1000	1	10^5	14.50	29.53	0.9869
N/m^2	0.01	10^{-5}	1	1.45×10^{-4}	2.95×10^{-4}	9.87×10^{-6}
lb/in^2	68.95	0.06895	6895	1	2.036	0.06805
in Hg	33.86	0.03386	3386	0.4912	1	0.03342
atm	1013	1.013	1.013×10^5	14.70	29.92	1

Velocity

SI units: mm/s, m/s, km/h
Imperial units: ft/s, mile/h

	mm/s	m/s	km/h	ft/s	mile/h
mm/s	1	10^{-3}	3.6×10^{-3}	3.28×10^{-3}	2.24×10^{-3}
m/s	1000	1	3.6	3.28084	2.23694
km/h	277.778	0.277778	1	0.911344	0.621371
ft/s	304.8	0.3048	1.09728	1	0.681818
mile/h	447.04	0.44704	1.609344	1.46667	1

Color Codes for Resistors and Capacitors

The British Standards Institution has defined two standards for the marking of values on components, a color-code system for resistors and a character–number system for both resistors and capacitors. The reader is referred to the following standard for further information: BS 1852 (1975) *Resistor and Capacitor Marking*.

Colour-code system for resistors

The value and tolerance of resistors is defined by a set of either four or five colored bands. These are displaced toward one end of the resistor with band one defined as that closest to the end of the resistor.

Four-band system

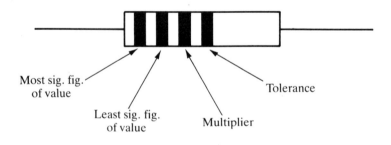

	Codes for bands 1–3				Code for band 4	
Black	0	Green	5	Brown	±	1%
Brown	1	Blue	6	Red	±	2%
Red	2	Purple	7	Gold	±	5%
Orange	3	Gray	8	Silver	±	10%
Yellow	4	White	9			

Five-band system (not often used)

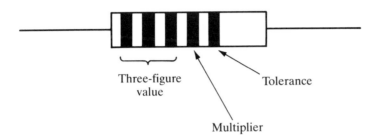

Three-figure value

Multiplier

Tolerance

		Codes for bands 1–4			Code for band 5	
Black	0	Green	5	Brown	±	1%
Brown	1	Blue	6	Red	±	2%
Red	2	Purple	7	Gold	±	5%
Orange	3	Gray	8	Silver	±	10%
Yellow	4	White	9			

Example 1

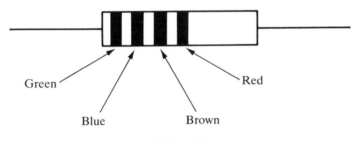

Green

Blue

Brown

Red

$$560\Omega \pm 2\%$$

Example 2 (not common)

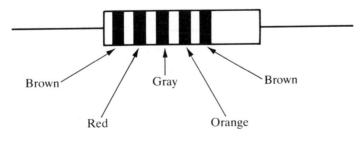

Brown

Red

Gray

Orange

Brown

$$128\,000 = 128\text{K}\Omega \pm 1\%$$

Character—number system for resistors and capacitors

This code uses two, three or four figures plus one letter to indicate value. Tolerance is indicated by a letter following the value code. Date of manufacture is sometimes indicated by a separate code.

Resistors

The value is defined by two, three or four figures.

The decimal point is represented by a letter which also acts as a multiplier: R, K, M, G and T define respectively $\times 1$, $\times 10^3$, $\times 10^6$, $\times 10^9$ and $\times 10^{12}$. For example,

6M8	means	6.8×10^6, i.e. $6.8\,\text{M}\Omega$
59R04	means	$59.04\,\Omega$

The code letters for tolerance are as follows:

B	$\pm$ 0.1%	F	$\pm$ 1%	K	$\pm$ 10%
C	$\pm$ 0.25%	G	$\pm$ 2%	M	$\pm$ 20%
D	$\pm$ 0.5%	J	$\pm$ 5%	N	$\pm$ 30%

Date of manufacture is indicated by either a two- or four-character code. In the two-character system, the code is as follows:

Code	Meaning	Code	Meaning	Code	Meaning
X	1969	M	1980	1	January
A	1970	N	1981	2	February
B	1971	P	1982	3	March
C	1972	R	1983	4	April
D	1973	S	1984	5	May
E	1974	T	1985	6	June
F	1975	U	1986	7	July
H	1976	V	1987	8	August
J	1977	W	1988	9	September
K	1978	X	1989	O	October
L	1979			N	November
				D	December

For example, R8 means manufactured in August 1983.

In the four-character system, the code is as follows:

First two characters	Year of manufacture
Second two characters	Week of manufacture

For example, 8507 means week 7 of 1985.

Capacitors

The value is defined by a sequence of numbers, as for resistors, with a multiplier code letter representing the decimal point.

Code	Multiplier
P	10^{-12}
n	10^{-9}
μ (u)	10^{-6}
m	10^{-3}
F	1

For example,

p10	$0.1\,\text{pF}$
333p	$333\,\text{pF}$
15n	$15\,\text{nF}$

Tolerance is indicated by the same one-letter code as for resistors. For example,

333pK   means   333 pF capacitor with $\pm$ 10% tolerance

Thevenin's Theorem

Thevenin's theorem is extremely useful in the analysis of complex electrical circuits. It states that any network which has two accessible terminals A and B can be replaced, as far as its external behavior is concerned, by a single e.m.f. acting in series with a single resistance between A and B. The single equivalent e.m.f. is that e.m.f. which is measured across A and B when the circuit external to the network is disconnected. The single equivalent resistance is the resistance of the network when all current and voltage sources within it are reduced to zero. To calculate this internal resistance of the network, all current sources within it are treated as open circuits and all voltage sources as short circuits. The proof of Thevenin's theorem can be found in Skilling (1967).

Figure A3.1 shows part of a network consisting of a voltage source and four resistances. As far as its behavior external to the terminals A and B is concerned, this

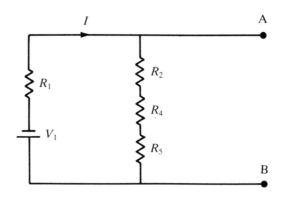

Figure A3.1

can be regarded as a single voltage source V_t and a single resistance R_t. Applying Thevenin's theorem, R_t is found first of all by treating V_1 as a short circuit, as shown in figure A3.2. This is simply two resistances, R_1 and $(R_2 + R_4 + R_5)$ in parallel. The equivalent resistance R_t is thus given by:

$$R_t = \frac{R_1(R_2 + R_4 + R_5)}{R_1 + R_2 + R_4 + R_5}$$

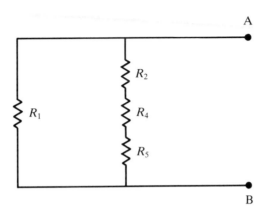

Figure A3.2

V_t is the voltage drop across AB. To calculate this, it is necessary to carry out an intermediate step of working out the current flowing, I. Referring to figure A3.1, this is given by:

$$I = \frac{V_1}{R_1 + R_2 + R_4 + R_5}$$

Now, V_t can be calculated from:

$$V_t = I(R_2 + R_4 + R_5)$$
$$= \frac{V_1(R_2 + R_4 + R_5)}{R_1 + R_2 + R_4 + R_5}$$

The network of figure A3.1 has thus been reduced to the simpler equivalent network shown in figure A3.3.

Figure A3.3

Let us now proceed to the typical network problem of calculating the current flowing in the resistor R_3 of figure A3.4. R_3 can be regarded as an external circuit or load on the rest of the network consisting of V_1, R_1, R_2, R_4 and R_5, as shown in figure

Figure A3.4

A3.5. This network of V_1, R_1, R_2, R_4 and R_5 is that shown in figure A3.6. This can be rearranged to the network shown in figure A3.1, which is equivalent to the single voltage source and resistance, V_t and R_t, calculated above. The whole circuit is then

Figure A3.5

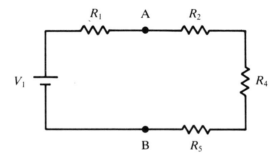

Figure A3.6

equivalent to that shown in figure A3.7, and the current flowing through R_3 can be written as:

$$I_{AB} = \frac{V_t}{R_t + R_3}$$

Figure A3.7

Thevenin's theorem can be applied successively to solve ladder networks of the form shown in figure A3.8. Suppose in this network that it is required to calculate the current flowing in branch XY.

Figure A3.8

The first step is to imagine two terminals in the circuit A and B and regard the network to the right of AB as a load on the circuit to the left of AB. The circuit to the left of AB can be reduced to a single equivalent voltage source, E_{AB}, and resistance, R_{AB}, by Thevenin's theorem. If the 50 V source is replaced by its zero internal resistance (i.e. by a short circuit), then R_{AB} is given by:

$$\frac{1}{R_{AB}} = \frac{1}{100} + \frac{1}{2000} = \frac{2000 + 100}{200\,000}$$

Hence,

$$R_{AB} = 95.24 \ \Omega$$

When AB is open circuit, the current flowing round the loop to the left of AB is given by:

$$I = \frac{50}{100 + 2000}$$

Hence, E_{AB}, the open-circuit voltage across AB, is given by:

$$E_{AB} = I \times 2000 = 47.62 \text{ V}$$

We can now replace the circuit shown in figure A3.8 by the simpler equivalent circuit shown in figure A3.9.

Figure A3.9

The next stage is to apply an identical procedure to find an equivalent circuit consisting of voltage source, $E_{A'B'}$, and resistance, $R_{A'B'}$, for the network to the left of points A' and B' in figure A3.9:

$$\frac{1}{R_{A'B'}} = \frac{1}{R_{AB} + 150} + \frac{1}{1000} = \frac{1}{245.24} + \frac{1}{1000} = \frac{1245.24}{245\,240}$$

Hence

$$R_{A'B'} = 196.94 \ \Omega$$

$$E_{A'B'} = \frac{1000}{R_{AB} + 150 + 1000} E_{AB} = 38.24 \text{ V}$$

The circuit can now be represented in the yet simpler form shown in figure A3.10. Proceeding as before to find an equivalent voltage source and resistance, $E_{A''B''}$ and $R_{A''B''}$, for the circuit to the left of A''B'' in figure A3.10:

$$\frac{1}{R_{A''B''}} = \frac{1}{R_{A'B'} + 250} + \frac{1}{500} = \frac{500 + 446.94}{223\,470}$$

Hence

$$R_{A''B''} = 235.99 \ \Omega$$

$$E_{A''B''} = \frac{500}{R_{A'B'} + 250 + 500} E_{A'B'} = 20.19 \text{ V}$$

Figure A3.10

The circuit has now been reduced to the form shown in figure A3.11, where the current through branch XY can be calculated simply as:

$$I_{XY} = \frac{E_{A''B''}}{R_{A''B''} + 300 + 200} = \frac{20.19}{735.99} = 27.43\,\text{mA}$$

Figure A3.11

Thermocouple Tables

Type E:	chromel–constantan
Type J:	iron–constantan
Type K:	chromel–alumel
Type N:	nicrosil–nisil
Type S:	platinum/10% rhodium–platinum
Type T:	copper–constantan

Temp. (°C)	Type E	Type J	Type K	Type N	Type S	Type T
−270	−9.834		−6.458	−4.345		
−260	−9.795		−6.441	−4.336		
−250	−9.719		−6.404	−4.313		
−240	−9.604		−6.344	−4.277		−6.105
−230	−9.456		−6.262	−4.227		−6.003
−220	−9.274		−6.158	−4.162		−5.891
−210	−9.063	−8.096	−6.035	−4.083		−5.753
−200	−8.824	−7.890	−5.891	−3.990		−5.603
−190	−8.561	−7.659	−5.730	−3.884		−5.438
−180	−8.273	−7.402	−5.550	−3.766		−5.261
−170	−7.963	−7.122	−5.354	−3.634		−5.070
−160	−7.631	−6.821	−5.141	−3.491		−4.865
−150	−7.279	−6.499	−4.912	−3.336		−4.648
−140	−6.907	−6.159	−4.669	−3.170		−4.419
−130	−6.516	−5.801	−4.410	−2.994		−4.177
−120	−6.107	−5.426	−4.138	−2.807		−3.923
−110	−5.680	−5.036	−3.852	−2.612		−3.656
−100	−5.237	−4.632	−3.553	−2.407		−3.378
−90	−4.777	−4.215	−3.242	−2.193		−3.089
−80	−4.301	−3.785	−2.920	−1.972		−2.788
−70	−3.811	−3.344	−2.586	−1.744		−2.475
−60	−3.306	−2.892	−2.243	−1.509		−2.152
−50	−2.787	−2.431	−1.889	−1.268	−0.236	−1.819
−40	−2.254	−1.960	−1.527	−1.023	−0.194	−1.475
−30	−1.709	−1.481	−1.156	−0.772	−0.150	−1.121
−20	−1.151	−0.995	−0.777	−0.518	−0.103	−0.757
−10	−0.581	−0.501	−0.392	−0.260	−0.053	−0.383

Temp. (°C)	Type E	Type J	Type K	Type N	Type S	Type T
0	0.000	0.000	0.000	0.000	0.000	0.000
10	0.591	0.507	0.397	0.261	0.055	0.391
20	1.192	1.019	0.798	0.525	0.113	0.789
30	1.801	1.536	1.203	0.793	0.173	1.196
40	2.419	2.058	1.611	1.064	0.235	1.611
50	3.047	2.585	2.022	1.339	0.299	2.035
60	3.683	3.115	2.436	1.619	0.365	2.467
70	4.329	3.649	2.850	1.902	0.432	2.908
80	4.983	4.186	3.266	2.188	0.502	3.357
90	5.646	4.725	3.681	2.479	0.573	3.813
100	6.317	5.268	4.095	2.774	0.645	4.277
110	6.996	5.812	4.508	3.072	0.719	4.749
120	7.683	6.359	4.919	3.374	0.795	5.227
130	8.377	6.907	5.327	3.679	0.872	5.712
140	9.078	7.457	5.733	3.988	0.950	6.204
150	9.787	8.008	6.137	4.301	1.029	6.702
160	10.501	8.560	6.539	4.617	1.109	7.207
170	11.222	9.113	6.939	4.936	1.190	7.718
180	11.949	9.667	7.338	5.258	1.273	8.235
190	12.681	10.222	7.737	5.584	1.356	8.757
200	13.419	10.777	8.137	5.912	1.440	9.286
210	14.161	11.332	8.537	6.243	1.525	9.820
220	14.909	11.887	8.938	6.577	1.611	10.360
230	15.661	12.442	9.341	6.914	1.698	10.905
240	16.417	12.998	9.745	7.254	1.785	11.456
250	17.178	13.553	10.151	7.596	1.873	12.011
260	17.942	14.108	10.560	7.940	1.962	12.572
270	18.710	14.663	10.969	8.287	2.051	13.137
280	19.481	15.217	11.381	8.636	2.141	13.707
290	20.256	15.771	11.793	8.987	2.232	14.281
300	21.033	16.325	12.207	9.340	2.323	14.860
310	21.814	16.879	12.623	9.695	2.414	15.443
320	22.597	17.432	13.039	10.053	2.506	16.030
330	23.383	17.984	13.456	10.412	2.599	16.621
340	24.171	18.537	13.874	10.772	2.692	17.217
350	24.961	19.089	14.292	11.135	2.786	17.816
360	25.754	19.640	14.712	11.499	2.880	18.420
370	26.549	20.192	15.132	11.865	2.974	19.027
380	27.345	20.743	15.552	12.233	3.069	19.638
390	28.143	21.295	15.974	12.602	3.164	20.252
400	28.943	21.846	16.395	12.972	3.260	20.869
410	29.744	22.397	16.818	13.344	3.356	
420	30.546	22.949	17.241	13.717	3.452	
430	31.350	23.501	17.664	14.091	3.549	
440	32.155	24.054	18.088	14.467	3.645	
450	32.960	24.607	18.513	14.844	3.743	
460	33.767	25.161	18.938	15.222	3.840	
470	34.574	25.716	19.363	15.601	3.938	

Temp. (°C)	*Type E*	*Type J*	*Type K*	*Type N*	*Type S*	*Type T*
480	35.382	26.272	19.788	15.981	4.036	
490	36.190	26.829	20.214	16.362	4.135	
500	36.999	27.388	20.640	16.744	4.234	
510	37.808	27.949	21.066	17.127	4.333	
520	38.617	28.511	21.493	17.511	4.432	
530	39.426	29.075	21.919	17.896	4.532	
540	40.236	29.642	22.346	18.282	4.632	
550	41.045	30.210	22.772	18.668	4.732	
560	41.853	30.782	23.198	19.055	4.832	
570	42.662	31.356	23.624	19.443	4.933	
580	43.470	31.933	24.050	19.831	5.034	
590	44.278	32.513	24.476	20.220	5.136	
600	45.085	33.096	24.902	20.609	5.237	
610	45.891	33.683	25.327	20.999	5.339	
620	46.697	34.273	25.751	21.390	5.442	
630	47.502	34.867	26.176	21.781	5.544	
640	48.306	35.464	26.599	22.172	5.648	
650	49.109	36.066	27.022	22.564	5.751	
660	49.911	36.671	27.445	22.956	5.855	
670	50.713	37.280	27.867	23.348	5.960	
680	51.513	37.893	28.288	23.740	6.064	
690	52.312	38.510	28.709	24.133	6.169	
700	53.110	39.130	29.128	24.526	6.274	
710	53.907	39.754	29.547	24.919	6.380	
720	54.703	40.382	29.965	25.312	6.486	
730	55.498	41.013	30.383	25.705	6.592	
740	56.291	41.647	30.799	26.098	6.699	
750	57.083	42.283	31.214	26.491	6.805	
760	57.873	42.922	31.629	26.885	6.913	
770	58.663	43.563	32.042	27.278	7.020	
780	59.451	44.207	32.455	27.671	7.128	
790	60.237	44.852	32.866	28.063	7.236	
800	61.022	45.498	33.277	28.456	7.345	
810	61.806	46.144	33.686	28.849	7.454	
820	62.588	46.790	34.095	29.241	7.563	
830	63.368	47.434	34.502	29.633	7.672	
840	64.147	48.076	34.908	30.025	7.782	
850	64.924	48.717	35.314	30.417	7.892	
860	65.700	49.354	35.718	30.808	8.003	
870	66.473	49.989	36.121	31.199	8.114	
880	67.245	50.621	36.524	31.590	8.225	
890	68.015	51.249	36.925	31.980	8.336	
900	68.783	51.875	37.325	32.370	8.448	
910	69.549	52.496	37.724	32.760	8.560	
920	70.313	53.115	38.122	33.149	8.673	
930	71.075	53.729	38.519	33.538	8.786	
940	71.835	54.341	38.915	33.926	8.899	

Temp. (°C)	Type E	Type J	Type K	Type N	Type S	Type T
950	72.593	54.949	39.310	34.315	9.012	
960	73.350	55.553	39.703	34.702	9.126	
970	74.104	56.154	40.096	35.089	9.240	
980	74.857	56.753	40.488	35.476	9.355	
990	75.608	57.349	40.879	35.862	9.470	
1000	76.357	57.942	41.269	36.248	9.585	
1010		58.533	41.657	36.633	9.700	
1020		59.121	42.045	37.018	9.816	
1030		59.708	42.432	37.402	9.932	
1040		60.293	42.817	37.786	10.048	
1050		60.877	43.202	38.169	10.165	
1060		61.458	43.585	38.552	10.282	
1070		62.040	43.968	38.934	10.400	
1080		62.619	44.349	39.315	10.517	
1090		63.199	44.729	39.696	10.635	
1100		63.777	45.108	40.076	10.754	
1110		64.355	45.486	40.456	10.872	
1120		64.933	45.863	40.835	10.991	
1130		65.510	46.238	41.213	11.110	
1140		66.087	46.612	41.590	11.229	
1150		66.664	46.985	41.966	11.348	
1160		67.240	47.356	42.342	11.467	
1170		67.815	47.726	42.717	11.587	
1180		68.389	48.095	43.091	11.707	
1190		68.963	48.462	43.464	11.827	
1200		69.536	48.828	43.836	11.947	
1210			49.192	44.207	12.067	
1220			49.555	44.577	12.188	
1230			49.916	44.947	12.308	
1240			50.276	45.315	12.429	
1250			50.633	45.682	12.550	
1260			50.990	46.048	12.671	
1270			51.344	46.413	12.792	
1280			51.697	46.777	12.913	
1290			52.049	47.140	13.034	
1300			52.398	47.502	13.155	
1310			52.747		13.276	
1320			53.093		13.397	
1330			53.438		13.519	
1340			53.782		13.640	
1350			54.125		13.761	
1360			54.467		13.883	
1370			54.807		14.004	
1380					14.125	
1390					14.247	
1400					14.368	
1410					14.489	

Temp. (°C)	Type E	Type J	Type K	Type N	Type S	Type T
1420					14.610	
1430					14.731	
1440					14.852	
1450					14.973	
1460					15.094	
1470					15.215	
1480					15.336	
1490					15.456	
1500					15.576	
1510					15.697	
1520					15.817	
1530					15.937	
1540					16.057	
1550					16.176	
1560					16.296	
1570					16.415	
1580					16.534	
1590					16.653	
1600					16.771	
1610					16.890	
1620					17.008	
1630					17.125	
1640					17.243	
1650					17.360	
1660					17.477	
1670					17.594	
1680					17.711	
1690					17.826	
1700					17.942	
1710					18.056	
1720					18.170	
1730					18.282	
1740					18.394	
1750					18.504	
1760					18.612	

Solutions to Exercises

Chapter 2

2.4 0.0175 mV/°C

2.5 9.8 Ω

2.6 (a)

Time (s)	Depth (m)	Temp. reading (°C)	Temp. error (°C)
0	0	20.0	0
100	50	19.716	0.216
200	100	19.245	0.245
300	150	18.749	0.249
400	200	18.250	0.250
500	250	17.750	0.250

(b) 10.25°C

Chapter 3

3.2 (a) 111001, 71, 39
   (b) 1100101, 145, 65
   (c) 10101111, 257, AF
   (d) 100000011, 403, 103
   (e) 1111100111, 1747, 3E7
   (f) 10011010010, 2322, 4D2

3.3

	One's compl.	Two's compl.
(a)	01010000	01010001
(b)	10001000	10001001
(c)	10011010	10011011
(d)	00101001	00101010
(e)	00010011	00010100

Chapter 4

4.1 378 mV

4.2 (a) $V_0 = V_i \left(\dfrac{R_u}{R_u + R_3} - \dfrac{R_1}{R_1 + R_2} \right)$

(b) 0.82 mV/°C

(c) Indicated temperature 102°C; error 2°C

4.3 24 V; 1.2 W

4.4 111 Ω; 5.0%

4.5 (a) $R_u = R_2 R_3 / R_1$; $L_u = R_2 R_3 C$

(b) $Q = \omega L_u / R_u = \omega R_2 R_3 C R_1 / R_2 R_3 = \omega C R_1$
Thus Q is proportional to R_1 at balance, i.e. a linear scale is required

4.6 (a) At balance

$$\frac{R_1 + j\omega L}{R_3} = \frac{R_2}{R_4 - j/\omega C}$$

By taking real and imaginary parts and manipulating:

$$L = R_1 / \omega^2 R_4 C \tag{A}$$

$$R_1 = \frac{R_2 R_3}{R_4 (1 + 1/\omega^2 R_4^2 C^2)}$$

Hence, at balance,

$$L = \frac{R_2 R_3 C}{1 + \omega^2 R_4^2 C^2} \tag{B}$$

(b) $Q = \omega L / R_1 = 1 / \omega R_4 C$

using equation (A) above. For large Q, $\omega^2 R_4^2 C^2 \ll 1$, and equation (B) above becomes

$$L = R_2 R_3 C$$

(c) 20 mH ·

Chapter 6

6.2 5.0%; 24 750 Ω

6.3 10.0%

Chapter 7

7.1   Mean 31.1; median 30.5; standard deviation 3.0

7.2   Mean 1.537; standard deviation 0.021; accuracy of mean value $= \pm0.007$, i.e. mean value $= 1.537\pm0.007$; accuracy improved by factor of 10

7.3 $\pm0.7\%$

7.4 $\pm4.7\%$

7.5 $\pm3\%$

7.6 $46.7\ \Omega\ \pm1.3\%$

7.8 $a=12.410$; $b=40.438$

Chapter 8

8.2   (a) Moving-coil (mean) 0.685 A; (b) Moving-iron (r.m.s.) 6.455 A

8.3   (a) 3.18 A; (b) 7.07 A, 6.37 A

8.4   (a) Mean 6.27 V; (b) r.m.s. 6.46 V

8.5   (a) 1.73 V; (b) 1.67 V; (c) 2.12 V

8.6   (a) Dynamometer (r.m.s.) 0.583 A, moving-coil (mean) 0.267 A; (b) 68 W

8.7   52.8 Hz

8.8   (b) $\omega_n = (K_s/J)^{1/2}$; $\beta = K_i/[2R(JK_s)]^{1/2}$; sensitivity $= K_i/(K_sR)$

       (d) 0.7

       (e) Typical bandwidth 100 Hz; maximum frequency 30 Hz

References and Further Reading

Chapter 3

Barney, G. C. 1985 *Intelligent Instruments*, Prentice-Hall

Chapter 4

Jones, C. 1962 *An Introduction to Advanced Electrical Engineering*, English Universities Press
Smith, R. J. 1976 *Circuits, Devices and Systems*, Wiley

Chapter 5

Blinchikoff, H. J. 1976 *Filtering in the Time and Frequency Domains*, Wiley
Hilburn, J. L. and Johnson, D. E. 1973 *Manual of Active Filter Design*, McGraw-Hill
Huelsman, L. P. 1970 *Active Filters: Lumped, Distributed, Integrated, Digital and Parametric*, McGraw-Hill
Lynn, P. A. 1973 *The Analysis and Processing of Signals*, Macmillan
Skilling, H. H. 1967 *Electrical Engineering Circuits*, Wiley
Williams, E. 1963 *Electric Filter Circuits*, Pitman

Chapter 7

Chatfield, C. 1983 *Statistics for Technology*, Chapman and Hall
Topping, J. 1960 *Errors of Observation and their Treatment*, Chapman and Hall

Chapter 8

Buckingham, H. and Price, E. M. 1966 *Principles of Electrical Measurements*, English Universities Press
Edwards, D. F. A. 1971 *Electronic Measurement Techniques*, Butterworths
Harris, F. K. 1952 *Electrical Measurements*, Wiley
Smith, R. J. 1976 *Circuits, Devices and Systems*, Wiley
Wolf, S. 1973 *Guide to Electronics Measurements and Laboratory Practice*, Prentice-Hall

Chapter 9

Doebelin, E. O. 1975 *Measurement Systems*, McGraw-Hill

Chapter 11

Doebelin, E. O. 1975 *Measurement Systems*, McGraw-Hill
Healey, M. 1975 *Principles of Automatic Control*, Hodder and Stoughton
IEE News 1985 No. 113, May

Chapter 12

Doebelin, E. O. 1975 *Measurement Systems*, McGraw-Hill

Chapter 15

Bentley, J. P. 1983 *Principles of Measurement Systems*, Longman

Chapter 16

Benedict, R. P. 1984 *Fundamentals of Temperature, Pressure and Flow Measurement*, Wiley
Bentley, J. P. 1983 *Principles of Measurement Systems*, Longman
Furness, R. A. and Heritage, J. E. 1986 'Commercially available flowmeters and future trends', *Measurement and Control*, **19** (No. 5), 25–35
Medlock, R. S. 1983 'The techniques of flow measurement', *Measurement and Control*, **16** (No. 1), 9–13
Medlock, R. S. 1985 'Cross-correlation flow measurement', *Measurement and Control*, **18** (No. 8), 293–8

Chapter 18

Baldwin, C. T. 1973 *Fundamentals of Electrical Measurements*, Prentice-Hall
Buckingham, H. and Price, E. M. 1966 *Principles of Electrical Measurements*, English Universities Press
Wolf, S. 1973 *Guide to Electronics Measurements and Laboratory Practice*, Prentice-Hall

Appendix 3

Skilling, H. H. 1967 *Electrical Engineering Circuits*, Wiley

INDEX